VIRTUAL ENGINEERING

VIRTUAL ENGINEERING

Edited by

J. CECIL, Ph.D.

School of Industrial Engineering and Management
Oklahoma State University

Momentum Press, LLC, New York

Virtual Engineering
Copyright © Momentum Press®, LLC, 2010

First published in 2010 by
Momentum Press®, LLC
222 East 46th Street, New York, N.Y. 10017
www.momentumpress.net

ISBN-13: 978-1-60650-097-2 (hard back, case bound)
ISBN-10: 1-60650-097-X   (hard back, case bound)

ISBN-13: 978-1-60650-099-6 (e-book)
ISBN-10: 1-60650-099-6   (e-book)

DOI forthcoming

Cover Design by Jonathan Pennell
Interior Design by Scribe, Inc. (www.scribenet.com)

First Edition: April 2010

10 9 8 7 6 5 4 3 2 1

Printed in Taiwan

Contents

Preface

Overview

The motivation underlying the publication of this book was to present a cross section of innovative practices and approaches in the broad area of Virtual Engineering. The intended audience includes researchers and practitioners in academia and industry who have interests in the following approaches and technologies:

- Virtual Reality (VR)–based prototyping techniques
- Collaborative Engineering
- Virtual Teaming approaches
- VR technology

Mechanical, Industrial, Manufacturing Engineering, and Computer Science students interested in learning more about practices and technologies in Virtual Engineering and Information Based Manufacturing (INBM) can use this book to develop an appreciation of innovative approaches and technologies within this field that is transforming collaborative engineering practices worldwide.

The Information Based Manufacturing Context

The contents of this book need to be read in the context of INBM. A brief note on INBM is necessary.

In 2001, I coined the term "Information Based Manufacturing" to reflect the emergence of this field. VR- and Internet-based technologies were two of the cornerstones of the Information Technology (IT) revolution (which was well underway at that time), which also served as powerful catalysts for the adoption of global collaborative engineering practices. INBM broadly refers to the study of concepts, principles, and approaches related to three core facets that impact the functioning of Virtual Enterprises (VE) in manufacturing and other domains. The (original) core facets included Modeling, Simulation, and Exchange of information as it pertains to product and process design activities. Each facet spans several areas, fields, or subfields of interest to engineers and computer scientists. A brief discussion of these three facets follows.

In "modeling" of information, the emphasis is on design and use of modeling languages to represent complex processes and facilitate the design of information-intensive engineering systems and approaches, including software systems for virtual manufacturing and other contexts. Modeling languages (such as the Unified Modeling Language [UML], the engineering Enterprise Modeling Language [eEML], etc.) are being widely used to develop complex process maps as well as to develop a formal understanding of distributed (and nondistributed) activities that subsequently become the basis for process design, improvement, and integration activities.

"Simulation" (in an INBM context) encompasses a range of topics including the study of the representation of product/process characteristics in VR-based environments, the use of virtual prototyping techniques to support product/process design tasks, as well as the investigation of visualization issues (such as the appearance of a target surface as a result of a specific process) that are essential to the creation of high-fidelity simulation environments.

"Exchange of information" includes a diverse range of areas and issues pertaining to the design of IT frameworks and Cyberinfrastructure to facilitate information exchange among distributed entities and resources in a VE. It also includes the use of telerobotic approaches, virtual team-based collaborations, as well as the design of semantic frameworks to support seamless information exchange across heterogeneous computing platforms.

The title of this book, *Virtual Engineering*, reflects the thematic nature of the eight chapters, which deal with virtual engineering issues (within the simulation and exchange facets). Six of these chapters (Chapters 1 through 6) deal with issues relevant to simulation and virtual prototyping. Two of the chapters (Chapters 7 and 8) deal with research issues that fall under the "exchange" facet. A brief overview of these chapters follows.

Chapter 1

In this chapter, a brief introduction to Virtual Prototyping (VP) is provided along with a review of selected virtual environments in emerging areas of microassembly and bioengineering. Some of the key challenges facing the adoption of VP practices are also discussed.

Chapter 2

This chapter introduces a new NonUniform Rational B-splines (NURBS)–based shape representation model for users to modify in a natural and intuitive manner. The authors delineate a NURBS modeling system that allows the designer to edit NURBS surfaces in real time using a glove called the ModelGlove.

Chapter 3

Maintainability issues and activities are often ignored but crucial aspects in the product development life cycle. The authors explore the use of VR technology in addressing such issues and propose a framework for product maintenance training.

Chapter 4

This chapter outlines the design and implementation of an advanced Virtual Assembly environment that supports the simulation of heavy machinery assembly using overhead cranes. Such an approach can facilitate the simulation of large-scale assembly scenarios (e.g., the assembly of large press machines), which can help evaluate assembly alternatives effectively.

Chapter 5

An innovative framework for haptic modeling and simulation of needle-insertion tasks in the context of medical education and training is discussed in this chapter. A client-server–based haptic framework is described to support multiple haptic devices that can function without inhibiting overall system performance.

Chapter 6

In this chapter, the authors elaborate on an Augmented Reality approach to create virtual exercise environments. These environments will enable people with disabilities to train and exercise virtually.

Chapter 7

With the advent of Internet and Internet2 technologies, the adoption of telerobotics and distributed manufacturing practices are becoming more commonplace. This chapter introduces recent results from the field of telerobotics that would be useful for virtual manufacturing, telesurgery, and other domains.

Chapter 8

In this chapter, the author discusses a case study that illustrates the barriers to effective collaboration among virtual teams. It also highlights the complex and multifaceted nature of collaboration challenges.

In today's globally distributed and IT-intensive engineering environment, it is essential for engineers and engineering students to keep abreast of innovative practices as well as established technologies. However (as discussed in Chapter 1 of this book), only a limited number of universities offer engineering programs that emphasize Virtual Engineering subjects. It is my hope that this book will underscore the need for more elaborate and focused discussions on the introduction of new courses as well as the creation of comprehensive educational programs in virtual engineering and INBM.

J. Cecil, Ph.D.
School of Industrial Engineering and Management
Oklahoma State University
Stillwater, Oklahoma
November 2009

Virtual Prototyping in Engineering

J. Cecil, Ph.D.

Center for Information-based Bioengineering and Manufacturing (CINBM),
School of Industrial Engineering, Oklahoma State University, Stillwater

Jeffrey Huber

Industrial Engineering, North Carolina State University, Raleigh

Abstract

Virtual Prototyping involves the creation of three-dimensional (3-D) models using Virtual Reality (VR) technology. In recent years, a range of engineering problems have been addressed through the creation of advanced Virtual Prototypes (VPs) that seek to provide a better understanding of a target product, process, or system. These VPs enable the earlier detection of infeasible product or process design ideas and facilitate collaborative approaches to complex problems by providing a more effective basis for communication among cross-functional teams. In this chapter, the key definitions and concepts are initially discussed, followed by a review of virtual prototyping research in emerging fields of engineering. It concludes with a summary of the key challenges in Virtual Prototyping that need to be addressed by future research.

Keywords: Virtual Prototyping, Virtual Reality, Virtual Engineering, Process Design

1   Introduction

In the research literature, there have been a range of definitions and descriptions for the term "virtual prototype" (VP). A more complete discussion of this term can be found in Cecil and Kanchanapiboon (2007). A VP can be described as a three-dimensional (3-D) computer model that seeks to mimic a target (or real-world) object, system, or environment using Virtual Reality (VR) technology (Cecil and Kanchanapiboon, 2007). This model can be a representation of a target object or system at various levels of abstraction.

The underlying process involved in the creation of a VP can be termed as "Virtual Prototyping." Figure 1 is an IDEF-0 diagram (at the A-0 level) modeling this process. The key inputs are customer requirements, specific preferences about the nature of the target VP to be created, and so on. The mechanisms include the Information Technology (IT) team, engineers, VR equipment, and associated software tools, including graphical libraries that are necessary to develop the VP. The controls include the project team's virtual prototyping expertise, knowledge about the target domain (for the VP being developed), and so on.

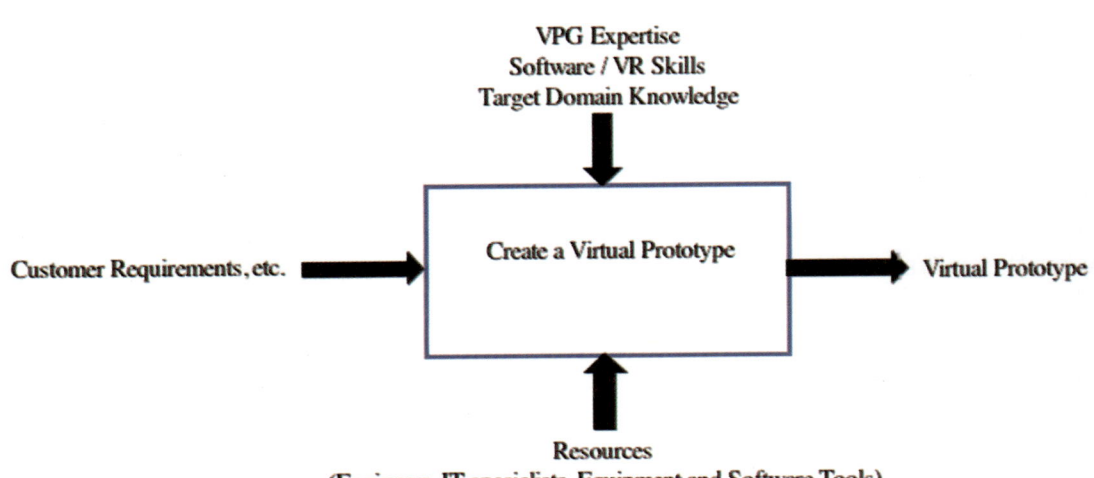

Figure 1.

A VP (in general) has certain characteristics that differentiate it from other models or prototypes. In Cecil and Kanchanapiboon (2007), these characteristics are listed and include the following:

1. *Appearance characteristics.* VPs must possess accurate geometry, topology, and appearance, reflecting characteristics of the target part, object, system, or environment.
2. *Simulation characteristics.* VPs should be capable of simulating engineering- or science-based characteristics, including behavior with real-time responses.
3. *Representation criteria.* A VP is a digital or computer-based representation.
4. *Interface criteria.* VPs must possess the ability to interface VR technology and graphics, including supporting semi-immersive or immersive applications.

A virtual prototyping approach provides several key benefits in engineering contexts. Its use facilitates better communication among engineering team members and enables the early detection of design issues from downstream process perspectives (including assembly feasibility, service feasibility, etc.). Consequently, it enables the reduction of overall lead time to manufacture a product as well as reduction in overall product development cost.

Zhang and Travis (2008) described the benefits of using a Virtual Assembly Environment (VAE) with multisensory capabilities that can provide support for stimulating a user's senses and provide a more realistic Virtual Environment (VE). The platform discussed is integrated with mono, visual, and 3-D auditory feedback to support investigation of the usefulness of a multisensory-enhanced VAE. The authors also provided the results of an evaluation related to the usefulness of this individual and integrated feedback. The results demonstrated that a VE with integrated feedback capabilities is more useful than individual or isolated feedback.

Vance, Su, and Seth (2008) outlined a system called SHARP (System for Haptic Assembly and Realistic Prototyping) to facilitate early detection of design and assembly problems in typical product development activities. A dual-handed haptic interface has been developed for virtual assembly and simulation applications that uses physics-based modeling. Users are provided with a platform to intuitively interact in the VAE, including the visualization of the dynamic behavior of rigid bodies; the immersive environment includes advanced physics behavior and collision-detection engines. As part of the validation activities, an array of complex computer-aided design (CAD) models have been assembled using SHARP.

In Xu, Jiang, Lu, and Wen (2008), a virtual workshop is discussed for 3-D modeling and rendering applications. The virtual workshop is composed of resources such as milling tools, lathes, and a machining center. 3ds Max and Virtools are used for 3-D modeling and rendering, respectively. The process outlined includes various tasks including creating the models, optimizing the number of triangles, script design, and subsequently creating the virtual workshop. The authors reiterate the key observation that the user's ability to interact with a natural "immersive" capability depends on the real-time performance of the VE. The paper concludes with a summary of the benefits of using a VR-based manufacturing environment.

In the following sections, a review of selected research efforts dealing with virtual prototyping in emerging areas is provided. A comprehensive discussion of virtual prototyping research in manufacturing and engineering can be found in Cecil and Kanchanapiboon (2007).

2 Virtual Prototyping Research

2.1 Virtual Prototyping in Emerging Domains

Micro Assembly, or Micro Devices Assembly (MDA), refers to the assembly of micron-sized parts or devices using manual, semiautomated, or fully automated methods. A micron is 10^{-6} meters. Several researchers have explored the use of VR in microassembly to help in assembly planning and comparison of gripping and assembly alternatives. Further, as it is difficult to work directly with parts in the scale of a few microns, a VE enables engineers to consider and study "what-if" scenarios virtually before interacting directly with physical microassembly resources. In the remaining sections of this chapter, a review of a limited number of research efforts in microassembly is discussed.

Monferrer and Bonyuet (2000) proposed a system to control robots in difficult or dangerous tasks using a VR-based framework. A set of guidelines is proposed to define an ideal user interface that would use VR to help an operator control an underwater robot. User interfaces are provided with video images and data such as depth and distance. A major drawback of the system is that the operator is expected to undergo extensive training. The authors identified three major categories of issues to developing a VR-based control system. These include (a) user interface issues, (b) technical issues, and (c) VR issues. The authors conclude that building an environment to control a physical system is an iterative process that can be improved with each new application task. This work highlighted the role of the human user in the design of VR-based collaborative environments.

Cecil and Gobinath (2005) outlined the creation of a VR-based simulation environment to support rapid assembly of microdevices. The virtual cell seeks to mimic the functioning of a physical microassembly work cell that is comprised of micropositioners, a part handling gripper (from Zyvex Corporation), cameras, and a workpiece supporting platen (WSP). The WSP is supported by a 3-degree-of-freedom (3-DOF) microtranslation stage constructed from three 1-DOF translation-stage model PI M-40-DG. The other components in the work cell are a Sony CCD-IDIS, an InfiniVar video microscope, a video monitor, and a computer. Varying voltage opens and closes the microgripper. The image from the video camera is processed through algorithms running in MATLAB, which relays the positions of the gripper, stage, and components to the VE.

The assembly-planning module provided two options to either generate an assembly plan manually or through a Genetic Algorithm (GA) approach. A user can manually enter an assembly sequence and 3-D path into a text file, which is read by the visualization module. If a user chooses to use a GA-based approach, then the GA-based sequence generator determines a near-optimal assembly sequence based on the position of the various bins containing the microdevices as well as the final destination of the micropins on the target part. Candidate assembly sequences, as well as path plans, can be compared using the VE (Figure 2). Alternately, the layout

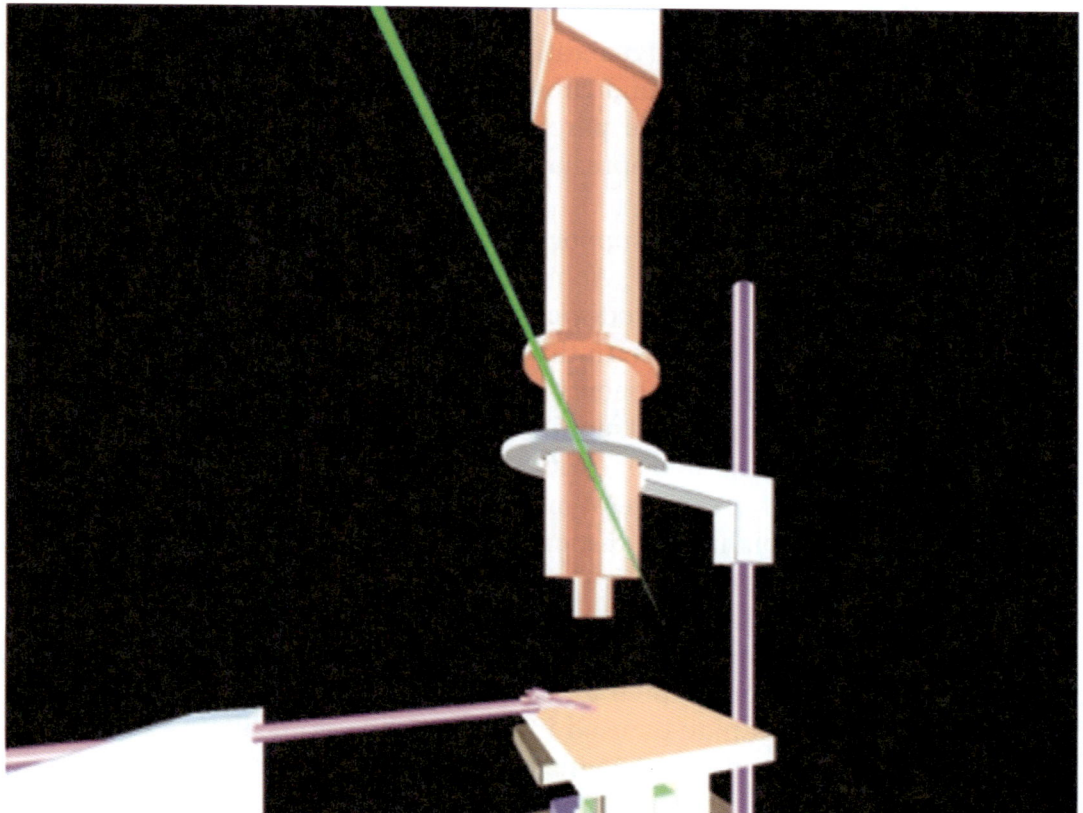

Figure 2. A view of the VE for microassembly.

of the bins (where the microdevices, such as pins, are held for a given assembly task) can be modified, and resulting assembly times associated with various layout options can also be compared. Subsequently, after a final plan is determined, the assembly instructions can be transmitted to the physical cell where the assembly tasks are completed. Feedback from cameras ensures satisfactory completion of the target microassembly tasks. The four other key modules involved in this process include the assembly-planning (or sequence-generating) module, the collision-detection module, the visualization manager, and the motion-generation module. The *collision-detection module* validated the assembly plan to ensure that no undesired collisions occur when the assembly resources (such as the gripper and micropositioners) are in operation. The visualization, or *VR world manager*, rendered all the data for user visualization. The *motion-generation module* receives candidate assembly sequences (generated within the GA-based module) as input and generates a collision-free 3-D path around possible obstacles in the assembly environment. The 3-D path is used to calculate the travel distance of the gripper when evaluating candidate assembly-sequencing alternatives.

The genetic operator works by randomly generating parent sequences (corresponding to assembly sequences), generating new child assembly sequences, determining the best child sequences, and subsequently generating the next generation of sequences from their respective parents. This process is continued until the new sequences do not show any appreciable difference in the traveling distance of the gripper.

A more advanced work cell to support microassembly tasks has also been developed at Oklahoma State University. A virtual assembly environment (Figure 3) to facilitate the rapid assembly of microdevices has been implemented using Coin3D (a 3-D graphics library) and C++ tools. In this advanced work cell, the physical assembly and positioning of microcomponents is achieved through an automated worktable, gripping unit, and camera and monitoring components. The worktable has two linear DOF in the x- and y-axis and one rotational DOF. The gripping system includes an innovative gripper that is capable of moving along the z-axis and has an angular DOF. The monitoring/visualization resources consist of two microscopes, two color video cameras, and illumination units. The microgripping unit is an innovative mechanism that allows an assortment of tweezer-shaped grippers to be used. This allows a comparison of various gripping surfaces and materials to be studied. This gripping mechanism is designed to be adapted to many different tweezer-like grippers. Many different types of tweezers may be used interchangeably to achieve a high level of adaptability, thus allowing different configurations to be assembled.

By interacting with the virtual and physical work cells (and using a range of micropositioners, microgrippers, cameras, and controllers), users can study

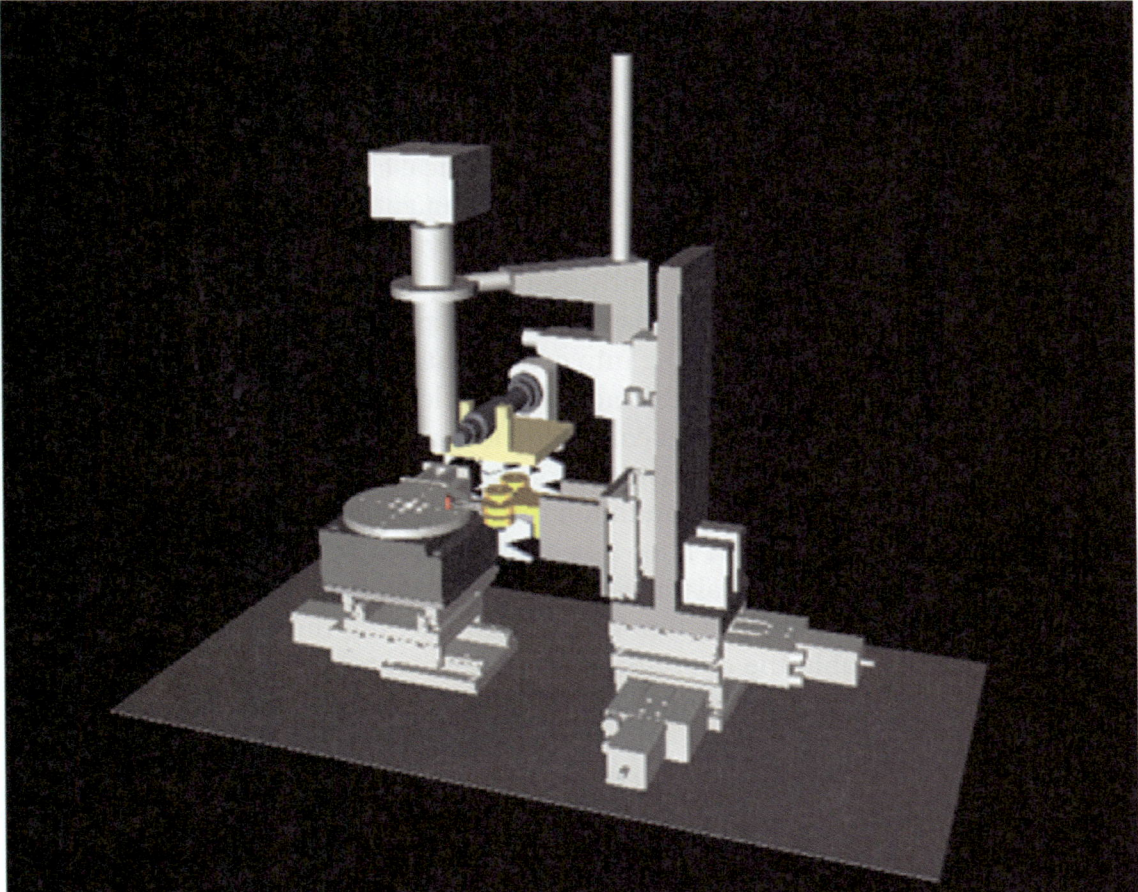

Figure 3. A virtual assembly environment for an advanced microassembly cell.

microassembly issues related to gripper-path planning and assembly. A range of pick and place tasks have been completed involving micron-sized pins and cams in the size range of 50 to 800 η.

Probst, Hürzeler, Borer, and Nelson (2009) outlined the use of a VR-based environment to control the assembly tasks using a microassembly system called the IRIS microassembly system V2. The physical workstation consists of a 6-DOF microassembly system including several microgrippers, a base unit, a top unit, cameras, and an illumination dome. The microassembly workstation has a graphical user interface (GUI) to enable users to assemble a microrobot for medical applications. The GUI contains a VR module that provides visual feedback of the real hardware. The VE is implemented using OpenSceneGraph. Before completing the physical assembly, a user can plan the various assembly tasks and then study them in a VE. Once the assembly task is studied to ensure feasibility, the instructions to complete the physical assembly are transmitted to the workstation controller. A collision-detection module is used to detect potential collisions during the assembly process.

Luo and Xiao (2006) discussed the assembly of optical fibers using haptic devices. The overall objective was to simulate an optical fiber–assembly process using VR with force feedback to facilitate the creation of an automated fiber-assembly process. The authors provided a discussion about the various problems in the manual fiber-assembly process and subsequently describe the modeling of micro- and nanoscale forces coming into play during assembly. The approach outlined the simulation of the fiber-assembly process, taking into consideration the interactive forces. Both noncontact- and contact-type forces models are relevant for micro- and nanoscale processes. The contact force model considers both adhesion and friction. The noncontact force models take into consideration van der Waals, electrostatic, gravitational, and capillary forces. The simulation results show the variation of noncontact forces and how they impact the fiber-assembly process. These force-based models are simulated in a VE with real-time haptic rendering using a PHANToM haptic device (with 6 DOF). Scaling factors are used to magnify these micro- and nano-forces forces that enable the operator to "feel" the contact and noncontact forces during the virtual fiber-assembly process.

Alex, Vikramaditya, and Nelson (1998) outlined an approach involving the integration of a VR Modeling Language (VRML)–based virtual microworld with visual servoing-micromanipulation strategies. The limitation of Micro Electro Mechanical Systems (MEMS) technologies in dealing with the complexities involved in the shape and properties of complicated parts creates the need for teleoperated micromanipulation and assembly. Such an integrated technique provides a vision-based feedback (using cameras), which helps in resolving the difficulties involved in the microdomain area. A user interacts with the virtual microworld that is represented by VRML. The manipulation of parts in this virtual world results in messages sent to the visual servoing agent in the form of a vector. The whole interaction between the virtual microworld, the user, and the servoing agent is enabled by the JAVA programming language. The dynamic and portable features of JAVA to operate on different platforms enable remote teleoperation over the Internet. The authors also discuss the overall framework of the implementation including the software and hardware components. The workstation has an optical microscope and two piezo-actuated nanopositioners.

Other researchers have also explored the use of VR to assist in microassembly tasks. These include Ferreira and Hamdi (2004); Bradley, Barmeshwar, and Joseph (1998); Sulzmann, Breguet, and Jacot (1995), Alex et al. (1998); and Cassier, Ferreira, and Hirai (2002).

2.2 VR in Bioengineering

VR-based surgical training environments have been the focus of several research investigations, including Megumi, Tomohiro, Hiroshi, Genichi, and Masashi (2006) and Bernhard, Alexander, Reinhard, and Dieter (2006). Other research efforts have explored the role of VR in bioengineering domains, such as Hamdi, Ferreira, Sharma, and Mavroidis (2008), who have investigated the use of VR for molecular dynamics–related simulations as it pertains to the prototyping of bionanorobots. In the remaining segments of this section, a review of VR-based simulation environments for bioengineering and other manufacturing domains is provided.

Luciano (2006) discussed the design of a prototype dental simulator for the training of periodontal procedures. The need for medical simulators has increased because of the limitations in existing conventional training methods. Using VR and haptics technology, the periodontical simulator allows students to learn to diagnose and treat periodontical diseases through interactive visualization of a virtual mouth enhanced with tactile sensations. The software for the haptics-based periodontal simulator is organized as two processes running concurrently (graphics rendering and haptics rendering, respectively). The graphics-rendering engine has been developed on top of Coin3D. The graphics-rendering process performs the stereo visualization of the 3-D mouth, dental instruments, and templates. The haptics rendering has been developed on top of GHOST (General Haptic Open Software Toolkit, a cross platform haptics library), which reads the current status of the haptic device, detects the collision between the virtual dental instruments and virtual mouth, and also computes the reaction forces to be applied. The efficient and novel integration between Coin3D and GHOST has worked to produce high-performance graphics and haptics rendering of relatively complex 3-D geometry for obtaining a realistic real-time user interaction. A subsequent experiment conducted by the Department of Periodontics at the University of Illinois at Chicago to validate the periodical simulator and its closeness to reality for training dental students has been proven successful.

Reitinger, Bornik, Beichel, and Schmalstieg (2006) described the development of a planning procedure for liver surgery using VR. Traditional surgery planning involves the use of volumetric information stored in a stack of intensity-based images like Computed Tomography (CT) scans where the surgeons can view them as 2-D images; the surgeons build their own visual/mental model of the liver, tumor, and vasculature using the 2-D images as references. Such an approach may lead to errors due to anatomical variability. An outline of the virtual liver surgery–planning system is presented, which includes three main stages: (a) image analysis, (b) segmentation refinement, and (c) treatment planning. Image analysis algorithms are used for automatic segmentation of liver, tumor, and vessels. Surgeons can subsequently correct any defects in the automatic segmentation by using segmentation refinement techniques. The treatment-planning tool is used to elaborate a detailed strategy for surgical intervention, which also includes an analysis of important quantitative indices, such as the volume of healthy liver tissue remaining after surgery. Validation of the segmentation-refinement tools with both artificial and real data sets were completed in a study involving medical students and physicians at the Medical University in Graz; evaluation of the treatment planning was also conducted using clinical data. Preliminary studies showed that the approach could be included in the clinical routine because of the benefits offered (interactive stereoscopic display, superior data quality, etc.)

Montgomery et al. (2006) implemented a real-time surgical-simulation system with soft-tissue modeling and network haptics. A general framework for surgical simulation that can support the requirements of many surgical simulation applications is developed, which encompasses a broad base of technological features with an emphasis on real-time performance. The system performs soft-tissue modeling, anatomy acquisition, simulation (suture modeling), collision detection, collision resolution, and sensor modeling. A number of displays are also supported, and other features such as voice input and output, real-time texture-mapped video input, stereo, and head-mounted display support are achieved.

3   Challenges and Directions for Future Research

While virtual prototyping has become an important part of product and process design (and engineering) activities on a global scale, there are several challenges that need to be addressed. Engineering enterprises that adopt VP strategies typically have to deal with the high cost of the equipment, software, and training. While the cost of the technology and peripherals has come down, fully immersive and semi-immersive environments are still expensive for most small businesses. Nonimmersive VR environments can be developed at a much lower cost using VRML 2.0 and freeware such as Coin3D. Some commercially available virtual-engineering tools (belonging to the nonimmersive category) can also be leased at a reasonable cost. However, any organization exploring the possible adoption of VP techniques and technologies has to study the impact of their use with care. Virtual Prototyping does not guarantee a reduction in overall product development costs. A pilot implementation involving use of VP approaches and technologies can enable the identification of potential problems. Successful implementations will involve a substantial amount of planning where software design principles and strategies are adopted. For example, process and functional models of the target implementation activities can be built in a structured manner using proven methods such as IDEF-0, IDEF-3, or eEML (Xavier, 2001).

Subsequently, the design of the VE can be formally captured using class, sequence, and collaboration diagrams (in the context of the Unified Modeling Language [UML]). Such formal approaches facilitate an efficient implementation involving the creation of target virtual-engineering environments. The use of these newly created VEs (or a COTS tool) can also be explicitly planned and modeled using a range of functional- or process-modeling approaches.

Virtual Prototyping approaches can play a significant role in the adoption of Concurrent Engineering (CE) approaches in manufacturing and engineering organizations. As they enable cross-functional teams with diverse perspectives to communicate more effectively, they can help with the earlier detection of infeasible product- and process-design ideas and plans; subsequently, they enable reduction in overall product development time and reduce product costs. The challenge in implementing such CE initiatives is to identify engineering personnel who are proficient in software design and implementation as well as CE concepts. In such situations, a steering team composed of multiple team members with backgrounds in IT, CE, and other engineering services can be formed to head the CE pilot initiative. Metrics need to be identified before the project is initiated; the performance of the pilot initiative in terms of reducing the product development time, overall costs, and so on can serve as the basis for continuous process improvement.

An important problem that is often overlooked is the lack of curriculum in undergraduate engineering programs that exposes students to VP techniques and technologies. While most mechanical, industrial, and manufacturing engineering programs introduce their students to Computer-Aided Design/Computer-Aided Manufacturing (CAD/CAM) technologies only a very limited number of institutions in the United States and worldwide have courses dealing with virtual prototyping and virtual engineering topics. While numerous computer science undergraduates are exposed to computer graphics topics and tools, very few academic programs provide a rigorous exposure to engineering problems and domains. Universities have to strive to develop educational programs at both the undergraduate and graduate levels that provide a balanced emphasis on engineering principles and VR technology. The lack of structured collaborations between software companies (who sell virtual prototyping tools) and universities (who are interested in teaching students the fundamentals of VP techniques) has contributed toward this scenario as well. The emergence of information based manufacturing (INBM) as an interdisciplinary engineering subject area is a first step toward the continuing evolution within engineering and computer science curricula worldwide. The growth of such educational programs at select universities is a limited response to the industry need for engineers who possess IT-oriented skills including virtual prototyping (Cecil, 2009). In today's IT-intensive engineering environment, the role of information has dramatically changed since the

1980s. Information is recognized as the key driver that propels and integrates various distributed life-cycle activities. The emphasis on the core facets of INBM (modeling, simulation, and exchange of information) for product and process design establish a strong foundation for the next generation of cybersystems that can support agile and flexible manufacturing strategies for a range of existing and emerging process domains (such as microassembly, nanoassembly, bioengineering, etc.).

Although the use of VPs has increased substantially over the past decade, it has not eliminated the practice of using physical prototypes. Physical prototypes, especially those manufactured by rapid prototyping techniques, continue to be widely used in industry. The primary reasons include the higher initial cost of VP technology as well as the difficulty in simulating advanced process characteristics and material attributes (such as surface roughness in machining operations, among others). Other reasons include the paucity of skilled engineers and software engineers who can develop robust VPs for target engineering domains. However, rapid prototyping has its own drawbacks. These include cost of physical prototyping, as well as the time involved in tooling changes and rework. A balanced use of both techniques and technologies will enable modern engineering organizations to respond to changing customer requirements.

While VR technology, in general, has developed rapidly over the past decade, one area that needs a greater degree of research effort and innovation is haptic technology. As VR approaches are adopted in medical and bioengineering fields, there is a need for substantial research to develop low-cost haptic devices that can be used to train surgeons and doctors virtually using VEs. The continued development of software toolkits to develop specialized training simulators is a key area not addressed by currently available commercial products. There is also a need for future research in exploring use of VR in micro-, nano-, and biosystems research.

4   Conclusion

This chapter provided an introduction to virtual prototyping as well as a review of selected research efforts in engineering. The main challenges and suggestions for future research in the field of virtual prototyping were also discussed.

Acknowledgment

Funding for the research activities that resulted in this chapter was obtained through grants (0965153 and 0951421) from the National Science Foundation (NSF) and a grant from Los Alamos National Laboratories. One of the NSF grants was an NSF

Research Experiences for Undergraduates (REU) Site grant awarded through the NSF REU program.

References

Alex, J., Vikramaditya, B., & Nelson, B. (1998, September). A virtual reality teleoperator interface for assembly of hybrid MEMS prototypes. *Proceedings of DETC '98 ASME Design Engineering Technical Conference.*

Bernhard, R., Alexander, B., Reinhard, B., & Dieter, S. (2006). Liver surgery planning using virtual reality. *IEEE Computer Society, 6,* 36–47.

Cassier, C., Ferreira, A., & Hirai, S. (2002, May). Combination of vision servoing techniques and VR-based simulation for semi-autonomous microassembly workstation. *Proceedings of the 2002 International Conference on Intelligent Robots and Systems.*

Cecil, J. (2009). Web site, http://iem.okstate.edu/People/Cecil.html

Cecil, J., & Gobinath, N. (2005). Development of a virtual and physical work cell to assemble micro-devices. *Robotics and Computer-Integrated Manufacturing, 21,* 431–441.

Cecil, J., & Kanchanapiboon, A. (2007, January). Virtual engineering approaches in product and process design. *International Journal of Advanced Manufacturing Technology, 31*(9–10), 846–850.

Ferreira, A., & Hamdi, M. (2004, September/October). Microassembly planning using physically based models in virtual environment. *Proceedings of the 2004 International Conference on Intelligent Robots and Systems,* IEEE-2004, *4,* 3369–3374.

Hamdi, M., Ferreira, A., Sharma, G., & Mavroidis, C. (2008). Prototyping bio-nanorobots using molecular dynamics simulation and virtual reality. *Microelectronics Journal, 39,* 190–201.

Luciano, C. (2006). *Haptics-based virtual reality periodontal training simulator.* Unpublished master's thesis, University of Illinois, Chicago, IL.

Luo, Q., & Xiao, J. (2006, October). Haptic simulation for micro/nano-scale optical fiber assembly. *Proceedings of the 2006 IEEE International Conference on Intelligent Robots and Systems,* 1353–1358.

Megumi, N., Tomohiro, K., Hiroshi, O., Genichi, S., & Masashi, K. (2006, September). Physics-based simulation fields for preoperative strategic planning. *Journal of Medical Systems, 30,* 371–380.

Monferrer, A. & Bonyuet, D. (2002, July). Cooperative robot teleoperation through virtual reality interfaces. *First International Symposium on Collaborative Information Visualization Environments,* London. Web site, http://vw.indiana.edu/cive03/

Montgomery, K., Bruyns, C., Brown, J., Sorkin, S., Mazzella, F., Thonier, G., et al. (2002, October). SPRING: A general framework for collaborative, real-time surgical simulation. *Medicine Meets Virtual Reality, 85,* 296–303.

Probst, M., Hürzeler, C., Borer, R., & Nelson, B. J. (2009). A microassembly system for the flexible assembly of hybrid robotic MEMS devices. *International Journal of Optomechatronics, 3*(2), 69–90.

Reitinger, B., Bornik, A., Beichel, R., & Schmalstieg, D. (2006). Liver surgery planning using virtual reality. Graz University of Technology. *IEEE Computer Graphics and Applications, 26*(6), 36–47.

Sulzmann, A., Breguet, J. M., & Jacot, J. (1995, October). Microvision system (MVS): A 3D computer graphic-based microrobot telemanipulation and position feedback by vision. *Proceeding of SPIE on Microrobotics and Mechanical Systems, 2593.*

Vance, M. J., Su, Hai-Jun, & Seth, A. (2008, December). Development of a dual-handed haptic assembly system: SHARP. *Journal of Computing and Information Science in Engineering, 8,* 1–8.

Xavier, B. (2001, October). Design of an Enterprise Modeling Language. VETI Report. Virtual Enterprise Technologies, Inc. (VETI). Las Cruces, NM.

Xu, L., Jiang, Z., Lu, H., & Wen, G. (2008). The application of virtual reality technology in constructing virtual machining workshop. *International Conference on Intelligent Robotics and Applications, Part I, 5314,* 479–487.

Zhang, Y., & Travis, A. R. L. (2008, April). Creation and evaluation of a multi-sensory virtual assembly environment. *International Journal of Automation and Computing, 5*(2), 163–173.

2 | Touch-Based Interactive NURBS Modeler Using a Force/Position Input Glove

Thenkurussi Kesavadas, Ph.D., Ameya Kamerkar, and Ajay Anand

Virtual Reality Laboratory, Department of Mechanical and Aerospace Engineering, University at Buffalo, State University of New York

Abstract

The creation of complex nonuniform rational B-spline (NURBS) surfaces is a tedious process because very few tools exist that allow a designer to design intuitively in real time. Standard input devices such as the mouse and the keyboard do not provide the designer with direct and easy methods for surface manipulation. We have developed a NURBS modeling system that allows the designer to edit NURBS surfaces in real time using a pressure-sensitive sculpting/molding input device designed to be worn as a glove called the ModelGlove. This input device is equipped with force and position sensors for quantifying touch and the intent of the designer. A virtual object deforms in a physically realistic manner in response to the user's direct manipulation of a hard or soft real physical object. The dynamic behavior of the NURBS model in response to the force and position input obtained from the ModelGlove produces highly natural shape variations. To demonstrate the viability of this technology, several complex 3-D shapes are demonstrated.

Keywords: interactive design, intuitive surface design, NURBS modeling, sculpting, virtual reality (VR) input device, virtual sculpting

1   Introduction and Motivation

Intuitive surface design and deformation have been extensively studied in both computer-aided design/computer-aided manufacturing (CAD/CAM) and computer graphics. Often after the surface or object has been created, further modifications are necessary. One common way to modify the shape of a free-form surface is to modify its control points one at a time. However, the modification process becomes tedious if the surface or object is composed of a large number of patches with many control points. Thus, interactive tools for manipulating a set of control points or sampled points are desirable in the case of complex sculptured surfaces.

Conceptual design is the initial stage of the design in which the essential form or shape of a product is created. During this stage, the specification of the product shape is not rigidly defined and the designer has some freedom in determining the features of the product. Although the conventional modeling approaches are ideal for certain applications, they tend to fall short of offering designers the flexible and unified ability to represent and interactively manipulate the surface models.

The methods used for free-form curve and surface modification in current CAD systems are still limited and nonintuitive. Many tools used for manipulation of free-form curves and surfaces are mainly based on changing the mathematical parameters, which requires the users to have an additional understanding of the mathematical principles involved. For example, while manipulating the NURBS surface, the user must know how changing the position of one control point will affect the shape of the surface. Similarly, the user must have an idea of the effect of *weights* on the surface. Generally, designers—and especially concept designers—prefer tools such as clay models, which more readily allow artistic and aesthetic design. Thus, the most natural tool for a designer is his or her hand. In this chapter, we have proposed a virtual surface modeling system composed of a tactile-based CAD modeling glove (ModelGlove) for capturing the motion of the user's hand, including pressure and the position of the fingers. The goal behind the development of such a system is to provide designers with a tool that will allow them to touch, push, and manipulate virtual objects just as they would clay models or sculptures. ModelGlove is based on a system developed by Mayrose, Chugh, and Kesavadas (2000a, 2000b) in our lab for measuring biomedical tissue properties.

SensAble Technologies' FreeForm modeling system uses PHANToM touch technology to allow sculptors and designers to model virtual objects on the computer using their sense of touch. It allows users to create 3-D design concepts and share them as 3-D models. It works like a 3-D mouse and provides real-time force and torque feedback to the user. However, this system is complex and relatively expensive. The ModelGlove is a simple, less expensive, yet powerful piece of hardware for manipulating NURBS models.

Using the ModelGlove as an input device, we introduce a new NURBS-based surface representation model for users to modify in a natural and intuitive manner. The new surface is generated by indirectly manipulating a set of control points based on the position and force applied using the ModelGlove. The method of control-point manipulation is indirect because the user applies the tool directly to the surface and the relevant control points are automatically modified so as to mimic the behavior of real clay. This is achieved through a displacement function. The displacement function is controlled by a set of key points that define the blending functions and a set of control vectors that are blended to form the final shape. The overall deformation of the parent surface can be viewed as the weighted average of the control vectors. The deformation of the surface is nominally based on physical laws. Through a computational physics simulation, the model responds dynamically to applied simulated forces in a natural and predictable way.

1.1 Prior Research

Research in geometric modeling has led to the development of many interactive and intuitive deformation methods for free-form curves and surfaces. NURBS have become the de facto industry standard for the representation, design, and data exchange of free-form geometric information. NURBS have been added to several international standards, and many commercial CAD packages include NURBS as a primitive modeling tool for designing free-form curves and surfaces. However, the NURBS paradigm is limited by the requirement that the surfaces are defined over rectangular domains, which leads to topological rectangular patches. Since its control points, weights, and knot sequences define a NURBS surface, modifications to these parameters produce a change in the shape of the surface.

Piegl and Tiller (1995) discussed a fundamental property of NURBS curves and surfaces called the cross ratio, which quantifies the push-pull effect of weights for NURBS curves. Piegl and Tiller (1995) and Welch and Witkin (1992) have also stated various shape-operator algorithms such as wrap, flatten, bend, stretch, twist, and taper. Au and Yuen (1995) proposed an approach for modifying the shape of NURBS curves by altering the weights and the location of control points simultaneously. The weights and control points are usually changed through user input from the keyboard and the mouse. Our approach builds on this methodology by incorporating the ModelGlove for a more intuitive feel of the sculpting procedure.

Free-form deformation (FFD; Hsu, Hughes, & Khaufman, 1992; Sederberg & Parry, 1986;) is a powerful NURBS-based technique for the deformation of free-form surfaces or volumes. It introduces a deformation model called the lattice that is represented by a trivariate volume regularly subdivided and defined by a 3-D array of control points. The object to be deformed is embedded inside the lattice. The transformation is applied to the lattice, and the embedded object is modified accordingly. But FFD is mainly used for global shape design and is not efficient for local surface design. Darrah, Kime, and Scoy (2002) have developed a convex hull approach for the selection of the nonplanar voxels. The PHANToM device is employed to select a region for manipulation. The algorithm uses the voxels within the region to define a convex hull. Once the voxels within the convex hull have been identified, the voxels can be modified easily.

Dachille, Qin, Kaufman, and El-Sana (1999) have directly manipulated a dynamic model of B-spline surfaces. Their method is physically based, as they evaluate the motion of surface nodes using a Lagrangian dynamics model. The PHANToM device has been used to operate tools with haptic feedback.

McDonnel, Qin, and Wlodarczyk (2001) have developed a sculpting system based on subdivision solids and physics-based modeling. The dynamic subdivision solids respond to the applied forces in a natural manner. However, also in this work, the force input is again provided through a PHANToM device.

Ehmann, Gregory, and Lin (2001) have developed the inTouch system for interactively editing and painting on a polygonal mesh using a PHANToM device. When touched by the PHANToM stylus, the meshes are divided into smaller ones to be displayed by the surface subdivision method. After the user has modified the mesh, he or she can interactively paint the mesh surface at the point of contact of the stylus with the surface.

Balakrishnan, Fitzmaurice, KurtenBach, and Singh (1999) have developed a device called ShapeTape for interactive NURBS curve and surface construction and manipulation. This device is a bend- and twist-sensitive strip that can be used intuitively with both hands. Bends and twists are measured at 6-cm intervals by fiber-optic bend sensors. By summing the bends and twists of the sensors along the tape, the shape of the tape relative to the first sensor can be reconstructed in real time. There is a one-to-one mapping between the tape and the NURBS curve.

Blaskó and Feiner (2003) have developed an input device composed of four pressure-sensitive linear strips. The user places each of the four fingers of one hand on a corresponding strip. The capacitance value associated with each strip is a function of the finger contact area, which in turn is dependent on the amount of pressure applied by the user. However, this system has not been used for CAD modeling but has been used as an advanced mouse to activate a multilevel 3-D menu system.

1.2 Design of the Input Device

The ModelGlove proposed in this work consists of a position sensor at the tip of one finger that senses the movements of this finger and a force sensor that reads the force data from the same fingertip (Figure 1). The position and force characteristics of the finger are tracked in real time and displayed graphically in the CAD modeling environment.

The magnetic position sensor, placed on the fingernail, tracks the movement of the finger in six degrees: namely, the translations along three axes and roll, pitch, and yaw about these axes. This sensor has a range of 30 in. The small size of this sensor allows the user to push deep into the nonmetallic object of study without interfering with its surface. The force sensor, which is located on the finger pad, collects data on the applied load from 0 to 25 lbs. The force sensor is .003 in. thick, which is similar to the thickness of most latex gloves worn by medical professionals. The thinness of the sensor allows the user to retain his or her sense of touch during the molding or sculpting process while simultaneously recording the force applied to the physical model. Both sensors can be programmed to collect data from 1 to 200 Hz, depending

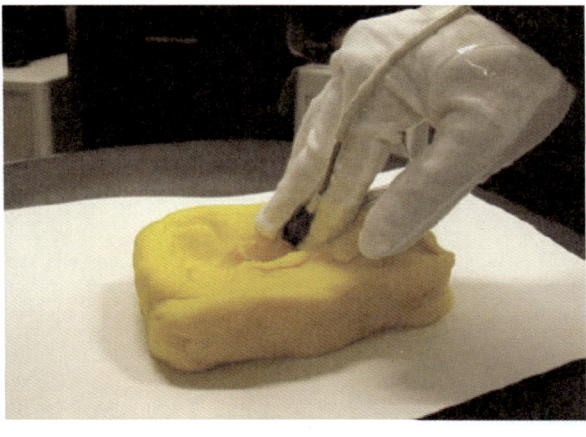
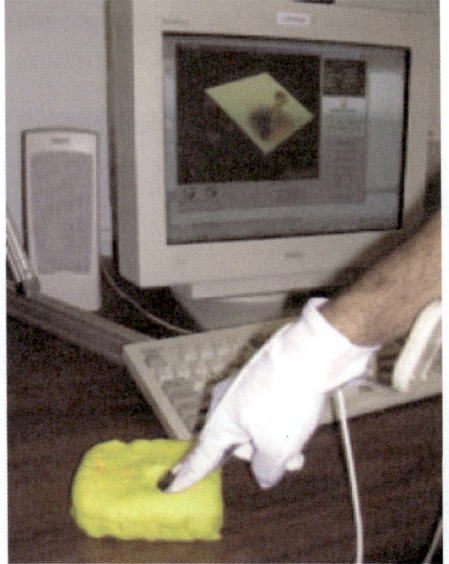

Figure 1. The ModelGlove and PC interface.

on the application. Current implementation uses just the index finger to carry out manipulative operations.

On the computer display, the user's finger is represented as a virtual tool. The position sensors sense the movement of the hand and interface those movements with the selected virtual tool. The force sensors capture the magnitude of the force exerted by the user. A choice of different tools is provided to allow intuitive and precise surface manipulation (Figure 2). Currently, four virtual tools are used: a sharp-point tool for making small deep holes, a medium-size ball for gauging or molding, a large-diameter tool for rough deformation of surfaces, and a suction tool to pull the surface out. The deformation field depends on the size of the tool tip and not on its exact shape (see Section 1.4).

To provide precise force input, the user is provided with several objects to touch, feel, and deform (e.g., a flat solid tablet, Play-Doh, spherical balls of different softness; Figure 1). Other physical objects may also be used based on the applications. When the user touches and applies pressure on one of these physical objects, the position of the fingertip, the applied force, and the time are collected and stored in a database. These data are then used to calculate the speed of fingertip motion. After the virtual surface has been created (as described in the next section), subsequent modifications can be implemented onto the generated surface by modifying the control points.

1.3 Sculpting on a NURBS Object

Nonuniform rational B-splines, or NURBS, are commonly used geometric primitives. NURBS allow the precise specification of free-form curves and surfaces as well as more traditional shapes, such as conics or quadrics.

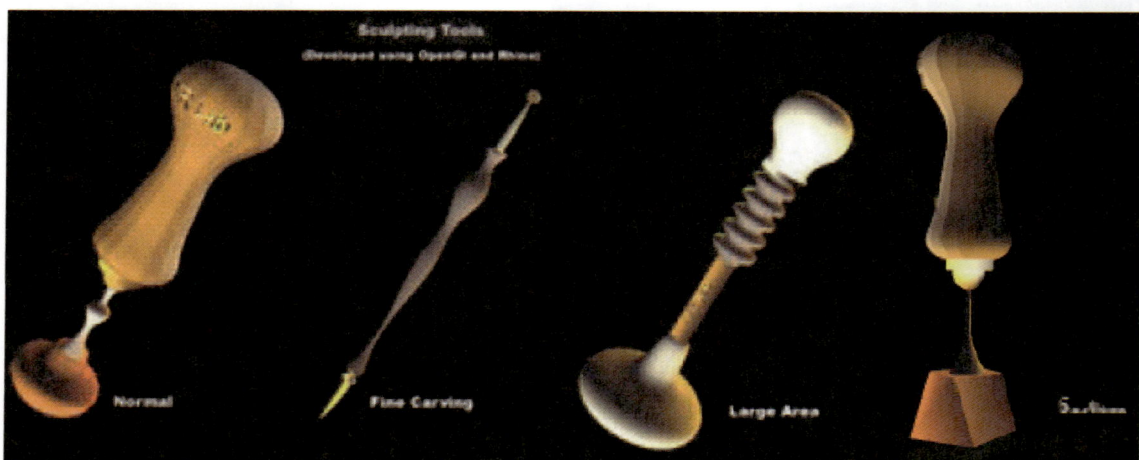

Figure 2. The four virtual surface editing tools.

A NURBS surface of degree (p, q) is defined by

$$S(u,v) = \frac{\sum_{i=0}^{m} \sum_{j=0}^{n} N_{i,p}(u) * N_{j,q}(v) * w_{i,j} * P_{i,j}}{\sum_{i=0}^{m} \sum_{j=0}^{n} N_{i,p}(u) * N_{j,q}(v) * w_{i,j}}, \tag{1}$$

where $N_{i,p}$ and $N_{j,q}$ are the B-spline basis functions, $P_{i,j}$ are the control points, and the weight $w_{i,j}$ of $P_{i,j}$ is the last ordinate of the homogeneous point $P_{i,j}^{w}$. Associated with the surface are two knot vectors, $U = \{u_0,u_1,K,u_r\}$ and $V = \{v_0,v_1,K,v_s\}$, where $r = n + p + 1$ and $s = m + q + 1$.

Changing a control point P_i or a weight w_i only affects the curve on the interval $u = (u_i, u_i + p + 1)$, which provides local control over the shape of the curve. Local control exists for surfaces as well. Modifying a control point $P_{i,j}$ or a weight $w_{i,j}$ affects only the portion of the surface in the rectangle $[u_i,u_i + p + 1] , [v_j,v_j + q + 1]$. Finally, curves and surfaces are infinitely differentiable on the interior of knot spans, and p-k times are differentiable at a knot of multiplicity k.

In the proposed modeling system, a surface representation is created that helps the user to modify an existing free-form surface (parent surface) in a natural and intuitive manner. The new surface is generated by adding a displacement function to the parent surface. The overall deformation of the parent surface can be viewed as the weighted average of the control vectors. The designer defines a point on the NURBS surface. Depending on his or her choice of tool, the force applied, and the position, the surface is deformed within the specified influence radius of the tool tip.

The magnitude of deformation of each control point is inversely proportional to its distance from the center of the tool tip and proportional to the total force applied. The lesser the distance from the center and the higher the force applied, the more the control point is displaced.

Figure 3 shows a cross-sectional view of the deformation process for a single B-spline curve. The control points 1, 2, 3, 4, and 5 lie within the influence radius of the tool tip R.

The distance d_i, for a control point i from the tool tip can be given as

$$d_i = \sqrt{\left(d_{ox} - d_{ix}\right)^2 + \left(d_{oy} - d_{iy}\right)^2 + \left(d_{oz} - d_{iz}\right)^2}. \tag{2}$$

As seen in Figure 3, the y component of the displacement increases with the decrease in proximity of the control point to the tool tip. The amount of deformation brought about by the tool varies with the influence radius R associated with each tool, as well

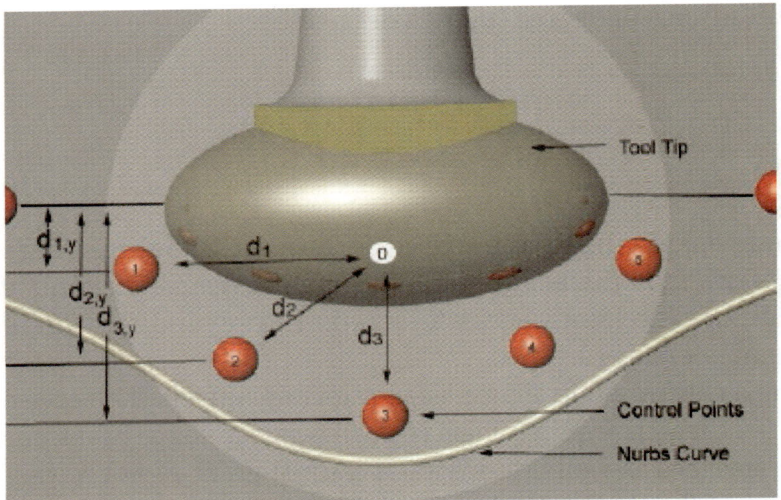

Figure 3. Control point deformation.

as the material properties assigned to the NURBS object. Currently, there are four virtual tools in use, as shown in Figure 2.

The user has a choice to select one of the three primitive shapes—namely, a sphere, a cube, or a cylinder—as the starting object. An additional pumpkin-like object was added for demonstration purpose. These objects differ from each other in several ways. Unlike a cube, which is a closed surface constructed by gluing different NURBS patches, each patch representing a single face, cylinder, sphere, and pumpkin is a single NURBS surface with free ends stitched together by overlapping points. The overlap points can be viewed as a pair of points belonging to opposite ends but always coincident. This ensures connectivity and continuity at the seam. This is illustrated in Figure 4. There is no limit on the maximum number of overlap points, but the minimum number depends on the order of the NURBS equation and the values of the knot points used. For example, if periodic knots are used, the number of overlapping points should be equal to the order of the NURBS equation, whereas nonperiodic knots require only one pair of points to overlap on the edges irrespective of the order.

A sphere and a pumpkin differ from the other two objects in the number of degrees of freedom of the control points. Their control points can be manipulated along all the three linear coordinates, whereas a cube and a cylinder deform only in orthogonal and radial directions, respectively. A smart-tool orientation technique has been implemented in which the tool orients itself normal to the sphere surface at the point of collision. This makes it easier for the user to orient the tool in the desired direction.

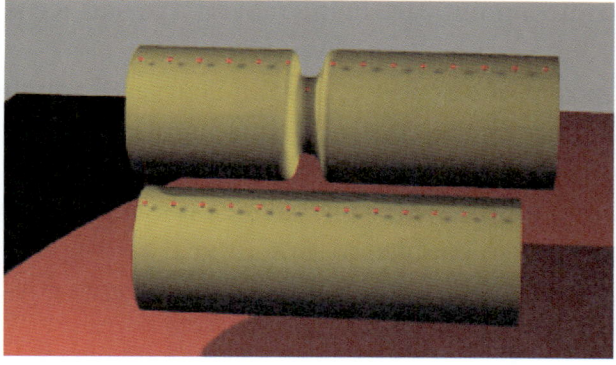 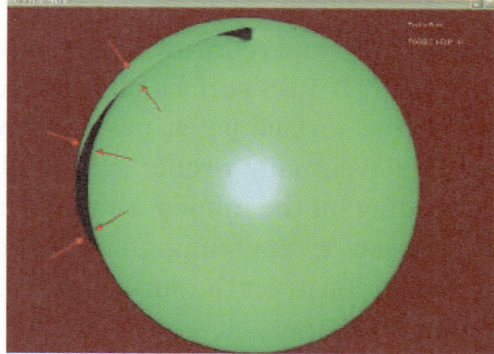

Figure 4. Volume conservation in a cylinder (left). Stitching free ends to form enclosed surface (right).

Volume conservation has been used for the cylinder in an attempt to simulate clay on a potter's wheel spinning at a high speed and subjected to radial deformation (Figure 4). It is done by shifting the control points toward the free end of the cylinder. The amount of shift is approximated by

$$\Delta l \approx 2d \frac{\Delta R}{R},$$

where d is the distance between any two adjacent control points along the length of the cylinder and ΔR is the change in the cross-sectional radius at the point of influence.

The NURBS used in all the objects was of the order of three. Round shapes like the sphere and the cylinder required standard manipulation of the weights of the control points. The pumpkin-like shape was obtained by increasing weights of alternate control points in U direction from a value of 1.0 to $\sqrt{2}$.

1.4 Editing the NURBS Surface With the ModelGlove

Editing a NURBS surface with the proposed ModelGlove input device requires that both the force and the position sensors be connected to the user's machine. To modify the control point using the ModelGlove, the user moves his or her hand to the desired location in the real world. The real-world object is mapped to the virtual clay on a 1:1 scale to provide an intuitive feel. When the user presses at the appropriate location on the physical object, the local region of the virtual object experiences the force exerted by the user. The size of the local region depends on the radius of the tool. The user can sculpt the NURBS object in the desired fashion based on his or her choice of tool. The virtual tool presses against the NURBS object and modifies it in the same fashion as a real object would. While the NURBS control points are being moved, the surface is recalculated and redrawn continuously.

The NURBS surface object is composed of a series of B-spline curves through which the surface patch passes. Any change in the control points associated with a curve eventually results in a local or global modification of the NURBS surface depending on the influence radius of the tool tip. To strike a balance between modeling accuracy and computational efficiency, we enabled a 16×16 grid of B-spline curves to construct the surface patch. Each of these curves has 16 control points governing its shape.

The equation of a generic B-spline curve is given as

$$\vec{C}(u) = \sum_{i=0}^{n} \left(B_{i,p}(u) * \vec{P}_i \right) , \tag{3}$$

where $B_{i,p}$ is the B-spline basis function for the curve of the order of n and degree $p - 1$. $\vec{P}_i$ is the ith control point composed of the x, y, and z coordinates associated with those points.

The modification of the NURBS surface is performed by modifying the location of the control point. The control point modification is affected by two actions of the ModelGlove. First, the position of the fingertip is obtained from the position sensor, and it is correlated to the nearest control point. The distance between the actual control point and the position of the fingertip is calculated. Second, two successive positions of the ModelGlove are used to compute a direction of the vector while the magnitude of the vector is obtained by the force sensor.

The amount of the change of the control point is proportional to the force. The force F applied by the designer using the ModelGlove can be given by the basic equation

$$F = kx + C\dot{x} , \tag{4}$$

where k is the stiffness, x is the displacement, C is the damping coefficient associated with the material, and $\dot{x}$ is the velocity imparted to the moving mass point.

For a nonelastic solid, the damping coefficient can be neglected. The actual displacement of the control points is governed by the direction of the vector of the force, which in turn is governed by the motion of the ModelGlove at the instant of force application (Figure 5).

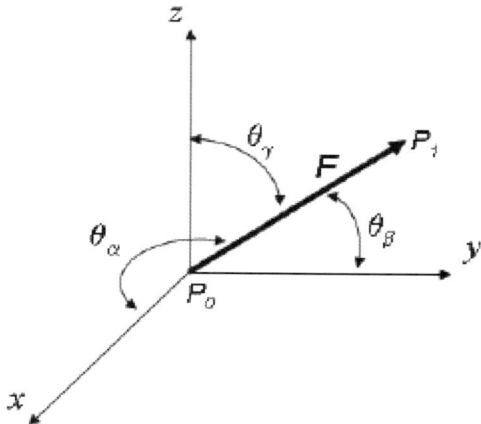

Figure 5. Direction of finger approach.

If θ_α, θ_β, and θ_γ are the angles made by the tool with the x, y, and z axes of the NURBS surface, the corresponding force components can be given as

$$F_x = Cos(\theta_\alpha) * |F|, \ F_y = Cos(\theta_\beta) * |F|, \ F_z = Cos(\theta_\gamma) * |F|. \tag{5a}$$

The displacement of the control point Pi can now be computed as

$$\partial P_{ix} = \frac{F_x}{k}, \ \partial P_{iy} = \frac{F_y}{k}, \ \partial P_{iz} = \frac{F_z}{k}, \tag{5b}$$

where k can be considered as a constant based on the physical object used for manipulation as described in Section 3. In our NURBS surface model, we have modified the conventional stiffness equation (Equation 5b) to form Equation 5c, which gives deformations very identical to deformations that result from tool-clay interaction (Figure 1).

$$\partial \vec{P}_i = \frac{\vec{F}}{k\left(r_i / r_o - 1\right)^m} a^{n_i}, \tag{5c}$$

$$\text{where } n_i = \sqrt{n_{i,u}^2 + n_{i,v}^2}. \tag{5d}$$

Here, ri is the distance between the tool tip center and the control point i. The representative tool size is ro. The deformation, $\partial\vec{P}$, attenuates as the topological distance between the control point i and the control point nearest to the tool increases. In Equation 5c, the parameter a ($0 < a < 1$) is the attenuation factor and ni is the topological distance. If ni,u and ni,v jumps are required in u and v directions (parametric surface coordinates), respectively, then the topological distance is given by Equation 5d. It can be shown that if m satisfies the inequality in Equation 5e, the distance between any two adjacent control points will be ensured to be less than radius of influence R. For this to be valid, the initial distance between adjacent control points should not exceed R.

$$m > \frac{\log\left(\dfrac{F_{max}\left(1-a\right)}{kR}\right)}{\log\left(2\right)} \tag{5e}$$

In other words, control points deform together in a group, and parameter a defines how loose or tight the group is. While R, ro, and m depend on the tool selected, k and a depend on the material selected. A particular set of these parameters can be stored in the form of tool and material templates. Updating the increments in the position of the control points in Equations 1 and 2, the new position of the control points is calculated in real time and the surface is modified accordingly.

The variations in the displacement behavior of the control points that lie within the influence radius of the sphere can be observed in the force-displacement graph (Figure 6). The tool is located closer to control point 4 than to points 3 and 5. As the magnitude of force is increased, the displacement of the control point, which is closest to the virtual finger, is observed to be more than the displacement of control points 3 or 5. The combined effect of the surface deformation hence is a function of the force applied and blending weighted functions obtained by the control points in the sphere of influence of the virtual tool.

1.4.1 Simulation Loop

Initially, all the control points are set to zero. Then the system runs in a loop and continuously updates the physical state of the sculpted object. The simulation loop traverses through the control points and computes the total internal forces acting on the points. External forces are queried from the ModelGlove attached to the computer.

Force Displacement Graph

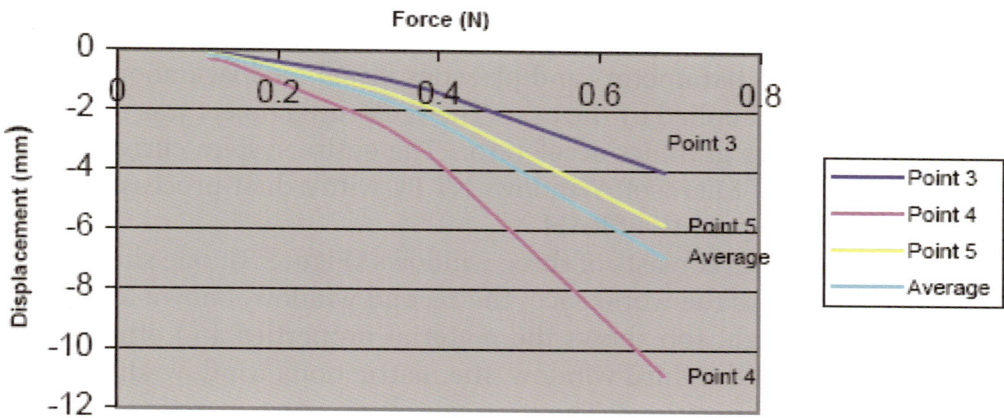

Figure 6. Force-displacement graph.

The acceleration and velocity of the control lattice are then computed in order to move the control lattice to its new position. The virtual sculpted surfaces can be updated at an interactive frame rate of 20+ frames per second, but a higher level of subdivision on the surface may degrade this performance. Surfaces can be edited in a wire frame mode or as a solid model surface. The control points for the surface are displayed as red spheres in Figure 7 (right), and an initial NURBS surface block before sculpting is shown in Figure 7 (left). Figure 7 (left) shows an initial NURBS surface block before sculpting.

Figure 7. The initial NURBS block (left) shows the control points (right).

1.4.2 The Software Implementation

The graphical user interface (GUI) for the software was developed in C++ on the OpenGL (Open Graphics Language) platform using GLUI libraries. The on-screen GUI controls sculpting parameters and provides visual feedback about the position and the force/position applied by the user. The sculptured object was rendered using OpenGL on a 3DS Labs graphics accelerator. The entire system currently runs efficiently even on an "old" Microsoft Windows NT PC with a dual-processor Pentium III with a 1-GHz CPU and 512 MB of RAM.

The visual interface consists of three windows (Figure 8): one shows a sculpted NURBS block, known as the workspace; the second window, known as the information window, at the right top shows the material properties and other parameters of the virtual clay; and the third window, the instructions window, displays current values of the force and positions. The user can also invoke an active graph window in place of the instructions window by pressing the "show graph" button. The new window, when invoked, shows a force-displacement graph that illustrates the characteristic behavior of the surface material in response to the forces applied by the user.

1.5 Model Interface to Commercial NURBS System

Initially, models designed using the proposed system could be saved and exported directly into a commercial CAD package Rhino model (.3dm). The file format was ideally suited for NURBS surface models because it stores the model information as discrete control points, knot points, degrees, and weights. This enabled easy data transfer to and from the Rhino modeler. The new IGES import and export module

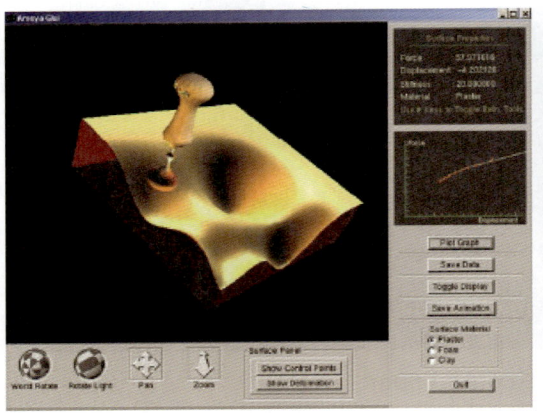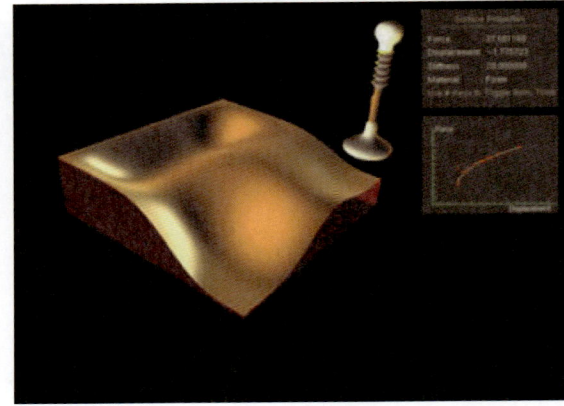

Figure 8. The software interface and the visual interface.

enables data transfer back and forth between our system and practically all other current modeling and animation packages without any data loss.

2   Results

Using our ModelGlove, several complex surfaces have been modeled (Figures 9 and 10). The surface blocks were created using three different virtual tools attached to the ModelGlove (Figure 2). The finished models were rendered in Rhino using the Flamingo Raytracer.

Figure 9. Object exported and rendered in Rhino.

Figure 10. A sample mould designed using our software and rendered in Rhino.

3   Conclusion

In this chapter, we have demonstrated a new NURBS modeling system along with a unique force-position input device that can be worn like a glove by a designer. This input device allows easy manipulation of surfaces by mimicking the process of an artist molding a clay object.

The NURBS-based model combined with ModelGlove technology makes our system computationally and economically less expensive than other systems. The results obtained using this system show that this system can be used to model fairly complex NURBS surfaces with little or no knowledge about modeling or computer programming. The proposed sculpting system has the potential of being a useful tool for artists and designers involved in modeling complex 3-D sculpted objects. User interaction with the CAD software using the simple, intuitive ModelGlove increases the realism of the design process and hence can also be used in virtual prototyping environments.

In the future, we plan to extend the ModelGlove prototype to include additional force and position sensors on the palm and two other fingers to provide more flexibility to the designer. A robust 3-D solid modeling package based on physically based models interfaced with the proposed input device is also being developed. Finally, we plan to carry out human-subject tests to study the effectiveness of our device over other devices currently used, such as simple mouse and keyboard inputs.

References

Au, C. K., & Yuen, M. M. F. (1995). Unified approach to NURBS curve shape modeling. *CAD, 27*(2), 85–93.

Balakrishnan, R., Fitzmaurice, G., KurtenBach, G., & Singh, K. (1999). Exploring interactive curve and surface manipulation using a bend and twist sensitive input strip. *ACM Symposium on Interactive 3D Graphics,* 111–118.

Blaskó, G., & Feiner, S. (2003). An extended menu navigation interface using multiple pressure-sensitive strips. *Seventh International Symposium on Wearable Computers,* 128–129.

Dachille, F., Qin, H., Kaufman, A., & El-Sana, J. (1999). Haptic sculpting of dynamic surfaces. In *Proceedings of the 1999 Symposium on Interactive 3D Graphics* (pp. 103–110). New York: ACM Press.

Darrah, M., Kime, A., & Scoy, F. (2002). A 3-D lasso tool for editing 3-D objects: Implemented using a haptics device. *Seventh Phantom Users Group Workshop,* 5–7.

Ehmann, S., Gregory, A., & Lin, M. (2001). A touch-enabled system for multiresolution modeling and 3D painting. *Journal of Visualization and Computer Animation, 12*(3), 145–158.

Hsu, W., Hughes, J., & Khaufman, H. (1992). Direct manipulation of free-form deformations. *Computer Graphics, SIGGRAPH '92,* Chicago, 177–184.

Mayrose, J., Chugh, K., & Kesavadas, T. (2000a). Material property determination of sub-surface objects in a viscoelastic environment. *Biomedical Sciences Instrumentation, 36*, 313–317.

Mayrose, J., Chugh, K., & Kesavadas, T. (2000b). *A non-invasive tool for quantitative measurement of soft tissue properties.* Paper presented at the World Congress on Medical Physics and Biomedical Engineering, Chicago. Proceedings on CD-ROM.

McDonnel, K., Qin, H., & Wlodarczyk, R. (2001). Virtual clay: A real-time sculpting system with haptic toolkits. *Proceedings of the 2001 Symposium on Interactive 3D Graphics.*

Piegl, L., & Tiller, W. (1995). *The NURBS Book.* New York: Springer-Verlag Berlin Heidelberg.

Sederberg, T. W., & Parry, S. R. (1986). Free-form deformation of solid geometric models. *SIGGRAPH '86, ACM Computer Graphics, 20*(4), 151–160.

Welch, W., & Witkin, A. (1992). Variational surface modeling. *Computer Graphics, 26*(2), 157–166.

3 | VR-Based Product Maintenance Training Systems

Q. Peng, Ph.D., T. Zhang, Y. Xie, and X. Kang
Department of Mechanical and Manufacturing Engineering,
University of Manitoba, Winnipeg

Abstract

Maintainability is an important criterion in product life-cycle management (PLM). Product maintenance requires experience and knowledge of product structure and operations. Using user-friendly interfaces to show product details and operations can significantly improve the understanding of a product and the efficiency of product maintenance. Virtual reality (VR) technology provides an effective tool for understanding product and maintenance training. Product structure and the operation process can be clearly shown in a VR-based user interface. In this research, a framework is proposed for product maintenance training using VR systems. The need for and implementation of the framework are discussed. Two applications are developed to demonstrate the feasibility of the proposed system.

Keywords: product life-cycle management (PLM), product maintainability, training, virtual reality (VR)

1   Introduction

Variety and multifunctionality make industrial products complex and unpredictable in some operations and maintenance situations. An operation may go wrong when interaction methods deviate from prescheduled operations. The maintainability measures a product or a piece of equipment to be repaired easily and efficiently when the product is damaged or the equipment breaks down. Product maintenance plays an important role in product life-cycle management (PLM) for any product's life expectancy. PLM considers products' sustainable development, maintainability, reliability, and recyclability. It requires various levels of detail and representations of product information.

Product maintenance consists of the diagnosis of the product's failures, product repair, installation of spare parts, and restoration of the product. Therefore, the total time spent in product maintenance includes time for diagnostics, time for obtainment of spare parts, time for parts removal and replacement, and time for system restoration. Product maintainability depends on the product's supportability, reliability-centered maintenance, integrated logistics support, and personnel training. A training system to understand product details and the operation process can reduce intervening time used for maintenance.

Considerations in product maintenance training include preventive strategies; methods for system testing and inspection; design for maintainability, critical components, and accessibility of components; and facilities and personnel for maintenance. Interactions between a product and operators are very important in product operation and maintenance. The quality of a system and product operation depends on the understandability of the product's structure, operations, and maintenance. Therefore, it is important to have proper support for the training of personnel in product maintenance to reduce repair time and the maintenance costs and to improve operational regularity. The recent combination of virtual reality (VR) technology with computer-aided design/computer-aided manufacturing (CAD/CAM) systems enhances the capacity of product development. Product developers can conduct product development processing with digital models. Products can be digitally evaluated in the early design stage, which eliminates unexpected errors and reduces the time and cost of product development.

VR systems provide a high degree of freedom for operational interactions. Improper operations can be simulated without incurring associated costs in terms of human injury and equipment repair. The complexity, size, and cost of the maintenance system justify the use of VR systems. Moreover, the time frame of development processing can be shortened by using the advantages and possibilities of VR technologies. A VR-based interface is proposed in this research for user-product interaction to increase the understandability and reliability of product maintenance. This chapter presents a VR system and implementation used for product maintenance training. Two applications have been developed. The first simulates the structure of a hydraulic cylinder used in a loading machine to guide the assembly and disassembly of components during product installation or repair. In the second application, the controllability of flows in an ultra-high-temperature (UHT) plant is monitored. The VR simulation of product maintenance reduces cost and time required in the operation training. The experience can be obtained from virtual models.

2 Related Research

Product variants have significantly increased the cost and complexity of PLM. Product maintenance is one of the most important elements in PLM (De Lit, Delchambre, & Henrioud, 2003), especially for a product with a long life cycle, such as an airplane. The airplane itself only takes about a third of its life-cycle cost, while two thirds of the cost are associated with maintenance (Waurzyniak, 2006). Maintenance operations are usually performed in a restricted space within a limited time span. The operator's movements are often constrained by surrounding components. Traditionally, product maintenance is carried out manually by engineers using their intuition and experience, as some operation processes are not easily formulated (Marcelino, Murray, & Fernando, 2003).

Because product maintenance is a dynamic and interactive process, the software tools must allow the user to operate a virtual maintenance process in dynamic environments. The current software tools are inappropriate for maintenance simulation (Marcelino et al., 2003). It is desirable to have a system for simulating realistic maintenance operations interactively. Virtual environments (VEs) have been used to assess maintenance operations before any physical prototype is available. The assessment of virtual models can reduce mistakes in physical operations. VR systems have also been effectively used in a variety of training purposes, including manufacturing facility design (Smith & Heim, 1999), surgical training applications (Low, Ilie, Welch, & Lastra, 2003), and training for the ship block spray painting (Yang, Lee, Shin, Hwang, & Son, 2007).

Research shows that vision counts for about 80% of the quantity of information received by the human brain during an error-detection task from an experimental study (Johansson & Ynnerman, 2008). A VR-based maintenance system allows visualization and manipulation of product structures, verification of processes, and maintenance process training. Much research has been done toward industrial applications of VR-based maintenance systems. Reuding and Meil (2004) investigated the performance of interactive graphics using an example of the ergonomics evaluation of a vehicle interior by comparing a user's performance in a VE with the performance in the real world. A simulation system was used to avoid the trial-and-error process on the real machine with costly materials (Puig, Perez-Vidal, & Tost, 2003). A VE-based visualization system was used with a collaborative product design system to achieve an ideal training environment for manufacturing industries. The VE as a learning tool makes product abstractions more concrete and thereby helps manufacturing companies internalize their learning (Jude, Vinesh, Raja, & Eyre, 2003).

VR has been approved as a cost-effective tool in manufacturing research (Chung & Peng, 2008). When integrated in manufacturing systems for product review and production evaluation, VR can enhance decision making for unseen problems in the manufacturing process (Peng & Yu, 2007). Cecil and Kanchanapiboon (2007) have presented a detailed literature review on virtual prototyping–related research in design and manufacturing. The benefits of VR have been translated into reduced overall product development cost and improved product quality. The engineer's response to different maintenance scenarios can be improved through the automated learning environment provided by VEs (Li, Khoo, & Tor, 2003). In general, there are three main requirements for a VR-based maintenance system: namely, visualization, interactivity, and free exploration. Experience-based knowledge has been linked to task performance to make assessments and evaluations less theoretical (Reuding & Meil, 2004). VR systems are faster to modify than real environments and thus easier to adjust to new conditions, and they additionally provide a hazard-free testing ground for participants.

A significant benefit of maintenance training using VR systems compared with traditional instruction is that it greatly enhances the interactive ability of the training process. However, because maintenance is a complicated process dealing with many elements of PLM, such as product design, manufacturing, assembly, disassembly, and process monitoring, existing software tools do not offer the function of simulating the mechanization in product maintenance (Puig et al., 2003). The process with conventional tools may encounter several limitations, such as loss of product structure, loss of the object's semantics, and loss of topological information (Wang & Li, 2006).

This chapter discusses the use of VEs as tools to enhance the usefulness of computer-aided product maintenance. Problems of existing VR tools will be identified. The need for development and further research will also be proposed. The applications developed show that visual displays provide a transparent, intuitive, and relatively inexpensive tool for the simulation of product maintenance. The simulation of operations allows maintenance to be well addressed in the training process. This reduces unforeseen problems that may progress through the product life cycle.

3 The Proposed Framework for VR-Based Product Maintenance

Functions designed for the proposed framework are to provide 3-D displays for navigating and manipulating the product and its components under maintenance and to integrate maintenance and visualization operations with a user-friendly graphical user interface (GUI). As each product has its performance criteria and interactions associated with the application, the ability to incorporate these criteria is essential to a VR system to improve on existing methods. It is important that the proposed system can be easily manipulated to enhance the training effectiveness of manipulations in virtual product maintenance environments. The system is designed to provide a trainee with sufficient information for maintenance alternatives to be evaluated, as shown in Figure 1. The VR user interface links all activities related to product maintenance. Product maintenance requires knowledge of product design, prevention, fault diagnostics, maintenance plans, and actions. The knowledge is built into a database or knowledge base related to the specific product. The product details and service requirements are used as the system input for the maintenance training. The actions of repair and maintenance are analyzed and simulated within the system. Often it is desirable to view dynamic events of product maintenance from different positions in a workshop or under different conditions.

There are three types of models built into a product maintenance system, as shown in Figure 2: (a) product-related models, (b) process-related models, and (c) dynamic models. Product-related models are the static entities of a product.

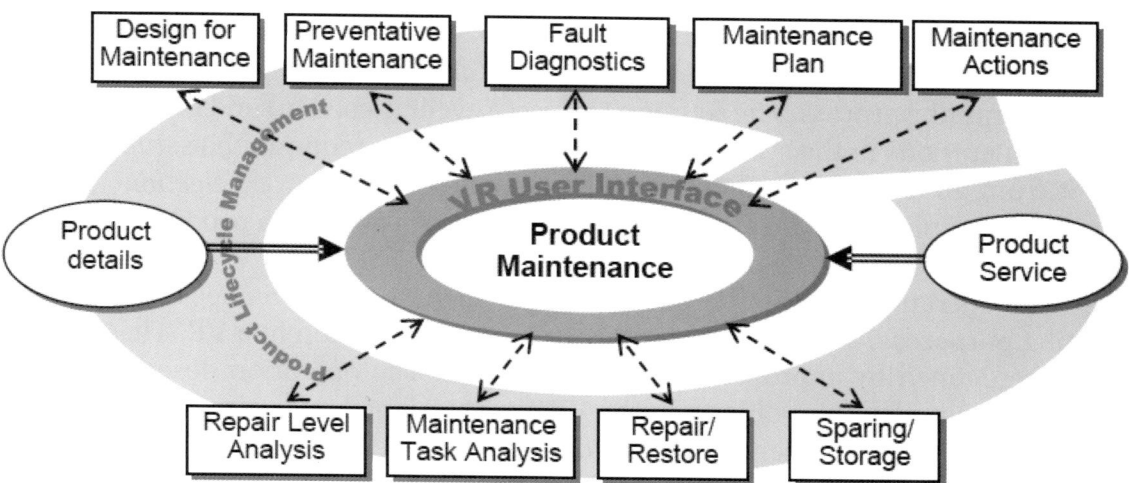

Figure 1. VR-based product maintenance for PLM.

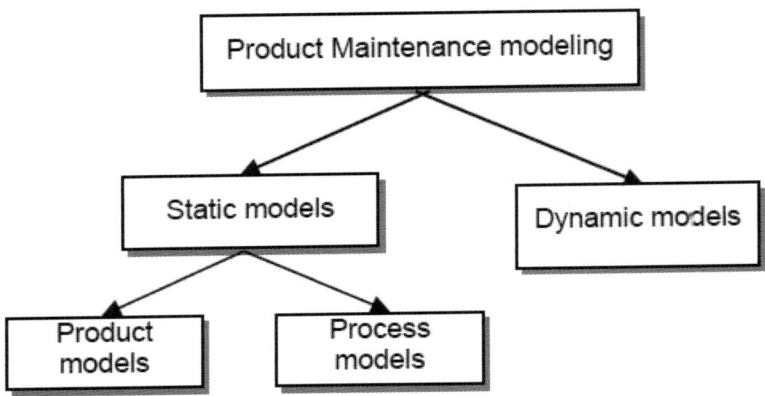

Figure 2. Models built into the product maintenance system.

Process-related models are the operation entities related to the product. Dynamic models are the entities that change during maintenance. A product assembly and components of the product are examples of product-related models. An operation scheme and a daily maintenance schedule are examples of process-related models. The failure of a product, a repair process, and the availability of storage are examples of dynamic models.

As shown in Figure 3, the proposed framework is implemented in three stages: (a) identification of the product and related components for product structure review, (b) identification of the product operation process, and (c) description of the maintenance operation process. All components in a product are defined by their relationships in an assembly. For example, a product may be reviewed with its components by an assembly or disassembly process with *productpart1( )*, *productpart2( )*, *productpart3( )*, *and productpart4( )*. A maintenance controller manipulates the training process. The *setupfortrainee( )* event is used to send messages to the related operations in the system. For example, a process requests the *setupthetool( )* event from an assembly and receives the response with the *toolselection( )* event. Another example is that it sends the *loadproductree( )* event to a product object and receives feedback from the *productstructure( )* event.

The operation description lists the tasks of the product model under specific training purposes. All operations have their own visualization in a VE. These operations are defined by appropriate motion elements. The operation description constitutes the simulation of objects in the VE given assigned operations. For example, when a product object receives the *startdisassembly( )* event, the assigned operation (*disassembly* of the product in this case) will be visualized. The parts' separation will

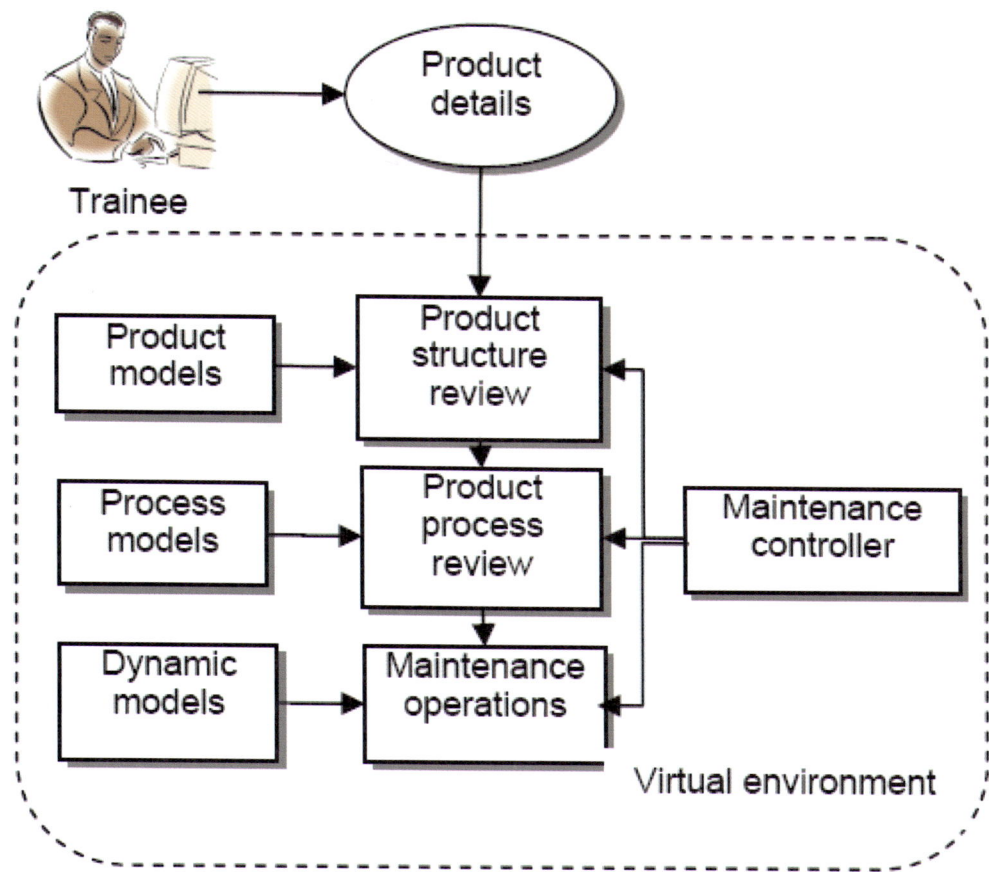

Figure 3. The procedure for product maintenance training.

be simulated, and tools used for the disassembly can also be seen gradually approaching and eventually executing the disassembly operation.

4   Implementation of the Proposed System

Because each product has its own feature, products may be designed with different tools and software. Especially for product maintenance, the process and operation may be different even for the same product. Importing product models created for VR worlds remains a challenge because of the variety of formats provided by various CAD systems. As the level of details in the representation of a product in CAD models is relatively low, it is necessary to create primitive elements with an external CAD program to represent product models for maintenance needs. The use of modeling tools supplied by third-party three-dimensional visualization (3DV) development environments is required.

The system implementation in this research is based on commercial tools available. Our work is the system integration and interface development. The existing toolkits provide most of the fundamental code needed to construct product models and visual interfaces. AutoCAD and Pro/ENGINEER are used as product modeling tools. The VR tool called EON Studio (EON, 2008) is used to create the user interface. EON Studio supports desktop or full immersive 3-D VEs for single or multiple users. The tool can be run on a single computer or use different computers for different aspects of VEs.

EON Studio also includes program prototypes for moving objects in the VE, constraining the rotation and translation of objects about axes, collision detection, and control of object attributes such as colors and lightings. After primitive product models are built by AutoCAD or Pro/ ENGINEER systems, the model can be imported into EON Studio. The controller is developed using EON Scripts to recognize and manage product models for operating product maintenance events in the VE. Representations of events are built for the various operations in maintenance training. As there are different maintenance needs for different products, the details of implemented systems will be described in the following section for each application.

5   Applications Developed for Product Maintenance

5.1 A Virtual Lab for Product Maintenance Training

Figure 4 shows the application of hydraulic cylinders in a loading machine. The hydraulic cylinder is a common component to be cared for in the product's lifetime. It is necessary to have knowledge about its internal structure, working processes, and the sequence of assembly and disassembly for product maintenance. Generally, it is difficult to demonstrate these clearly in a training course without experience and interaction with an appropriate model.

An application was developed based on the proposed VR system to provide an intuitive and easy way for quickly understanding the product for the purpose of maintenance training. Four virtual labs were developed based on the requirements of maintenance training, including the structure lab, operation lab, assembly lab, and disassembly lab.

The product's 3-D model is from the existing CAD model and is converted into EON format for maintenance operations. The model is linked with the navigation notes of the VR system to meet the needs of interactions between users and components.

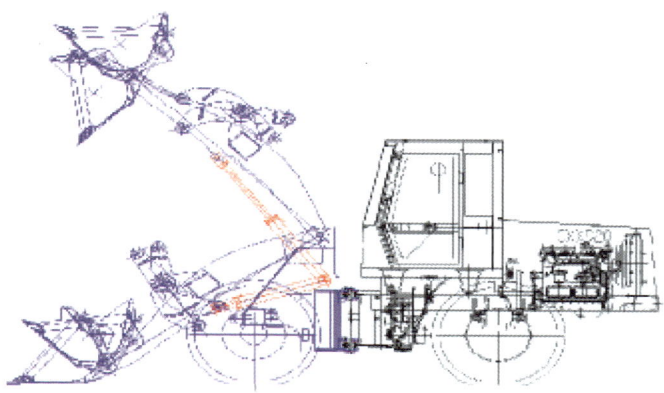

Figure 4. Hydraulic cylinders used in a loading machine.

Four camera frame notes are used to increase viewports. The initial positions and orientations of the cameras are set up for the model observation. In order to navigate the model in different views for different labs without interfering with each other, four ObjectNavLITE prototypes are developed into the system, including Object-NavLITE1, ObjectNavLITE2, ObjectNavLITE3, and ObjectNavLITE4. Their camera properties are set to Camera1, Camera2, Camera3, and Camera4, respectively. A MouseSensor node is added to capture the operation action. A Script node is developed to handle the LeftButtonDown event message from the MouseSensor. Figure 5 shows a flow chart of the message handler for the LeftButtonDown message sent from the MouseSensor.

Figure 6 shows the appearance of the simulation windows. Users can change the main view from one lab to another by clicking on the viewport located on the top of the simulation window. In the virtual assembly lab, initial positions of the components are different from other labs. All components can be predisassembled in this lab. Each lab has different functions. Some interactions, such as pressing a key or clicking on a component, are assigned for different training purposes. The instructions are also shown on the operation window of all labs during the operation.

5.2 UHT Operation and Maintenance Training

Ultra-high temperature (UHT) is used for the sterilization of liquid food before packaging, such as milk, juice, beer, wine, and other beverages. The liquid food is filled into presterilized containers in a sterile atmosphere after the sterilization. For example, milk is processed in this way using temperatures exceeding 135°C. The high temperature is required to kill spores in milk. The processing time has to be short, normally

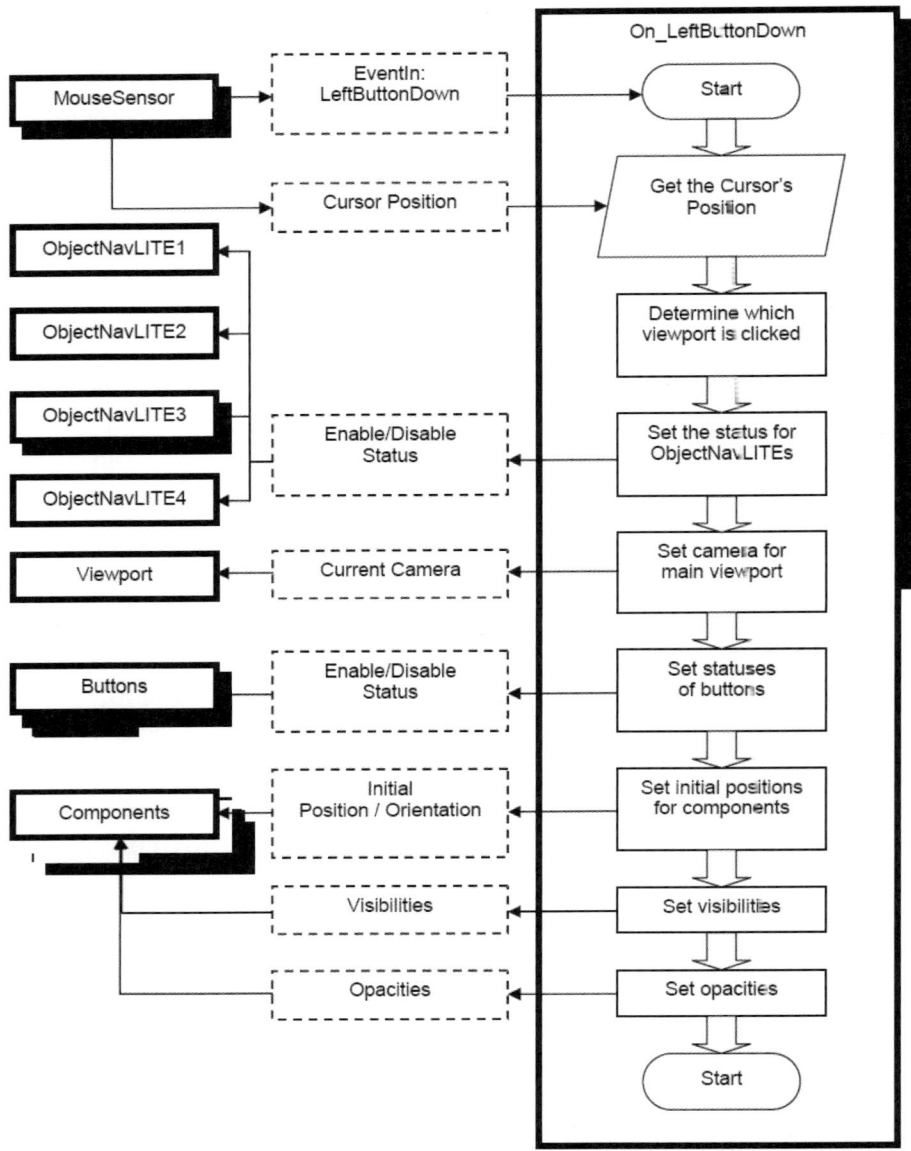

Figure 5. The flow chart of On_LeftButtonDown.

2 to 5 sec., for a continuous-flow operation. This treatment will keep the product safe during its shelf life, typically 6 to 9 months at room temperature until opened.

Normally, after a UHT plant is built for the customer, the UHT plant vendor will train operators after commissioning. The purpose of the training is to let the operators understand the working process of UHT. For example, a training course may cover the processes of heating milk and sterilization. Trainees can learn the operations of

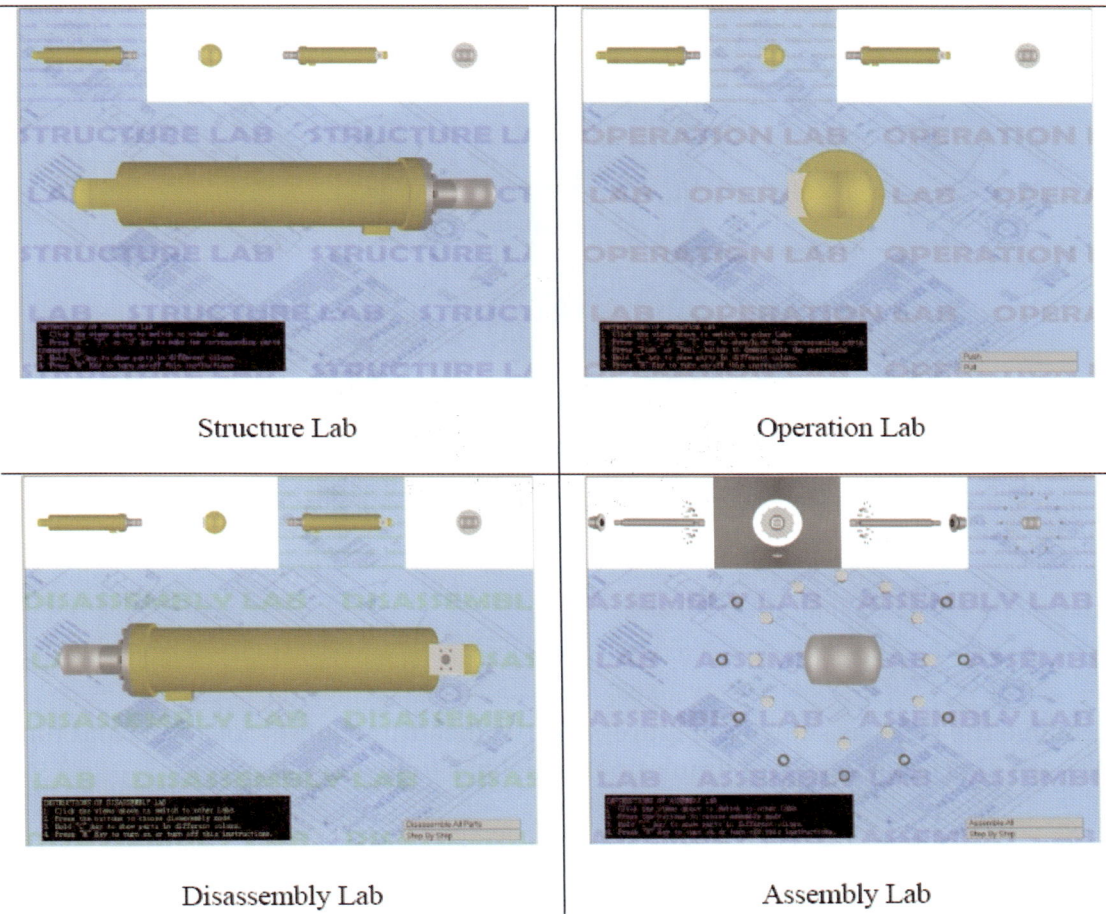

| Structure Lab | Operation Lab |
| Disassembly Lab | Assembly Lab |

Figure 6. Initial views of the virtual labs.

the UHT plant to ensure production safety. The flow directions of different liquid materials between different units are the basis of comprehension of such technology. During the training, these directions are normally indicated on a piping and instrumentation diagram (PID). The PID is a good tool to understand the working process. But when the operators face the real plant and the complex pipeline (as shown in Figure 7), it is hard for them to match the PID to what they see in reality.

Therefore, a 3-D virtual UHT plant model is built for improvement of operation training. The virtual model helps operators to understand each component's real location easily. Different colors, arrow movements, and view changes are used in the model to show the liquid flow direction through each pipe and unit. Five cameras are defined for five views to show the UHT plant model completely, including overview, view 1, view 2, view 3, and view 4. The trainee can click the relevant element button

Figure 7. An operation unit of UHT plant.

to change the view, as shown in Figure 8. In the meantime, units' locations and functions can also be shown. The simulation model has following functions:

- Locate the start and end points of liquid flows.
- Define flow direction changes and rotations of each moving segment.
- Define the flow moving time and direction changing time.
- Use a timer to control and connect all the flow movements and direction changes.

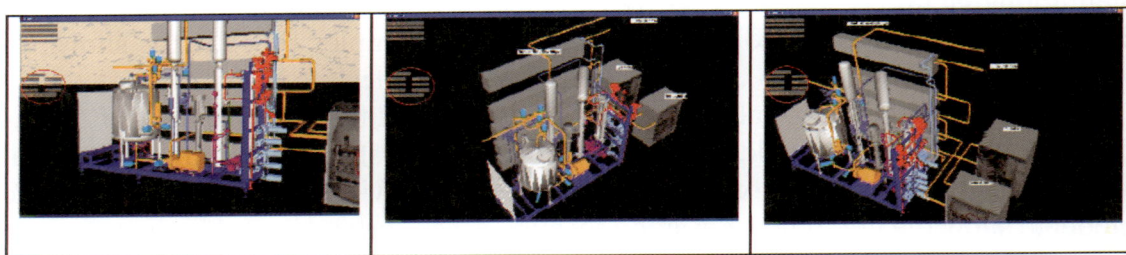

Figure 8. Different views of the model.

Generally, two loops exist inside the UHT unit: the product loop and the water loop. The simulation of the loops in the operation is indicated by the colored flow in the model, as shown in Figure 9. The user operation menu is shown on the left side of the window. When "product loop" is selected, the selected button will change color from gray to green for the action. A white arrow, indicating milk, will move through pipes and units, such as the balance tank, feeding pump, deaerator, homogenizer, heat exchangers, and filling machine. The trainee can change views following the arrow movement. When the loop finishes, the arrow will disappear. The hot water loop can be shown by a similar procedure. The three windows in Figure 9 are different views following the flows. Figure 10 shows the flow operation and view control in the simulation. Working loop selection and camera selection are decided simultaneously to get a better view of the operation. The display arrow is moved following the timers. In Figure 10, T_1, T_2, . . . T_n are different timers to count related flows indicated by arrows AA_1, AA_2, . . . AA_n for the display in the simulation. The arrow will disappear after the flow passes the observation point in the display window. C_1, C_2, . . . C_n are cameras to capture the different views V_1, V_2, . . . V_n in the operation simulation. There are several working phases for the UHT plant, including heating, sterilization, ready for production, production, and clean. Figure 11 shows an example of the simulation process for the production phase.

6 Discussions and Further Work

The following problems are identified from the applications developed:

- Existing product modeling systems do not provide functions related to product maintenance operations.

- Current VR tools only provide operations for model review and navigation.

- Most product CAD models cannot be directly used in existing commercial VR tools. The product has to be transferred with individual components of the

Figure 9. The flow simulation.

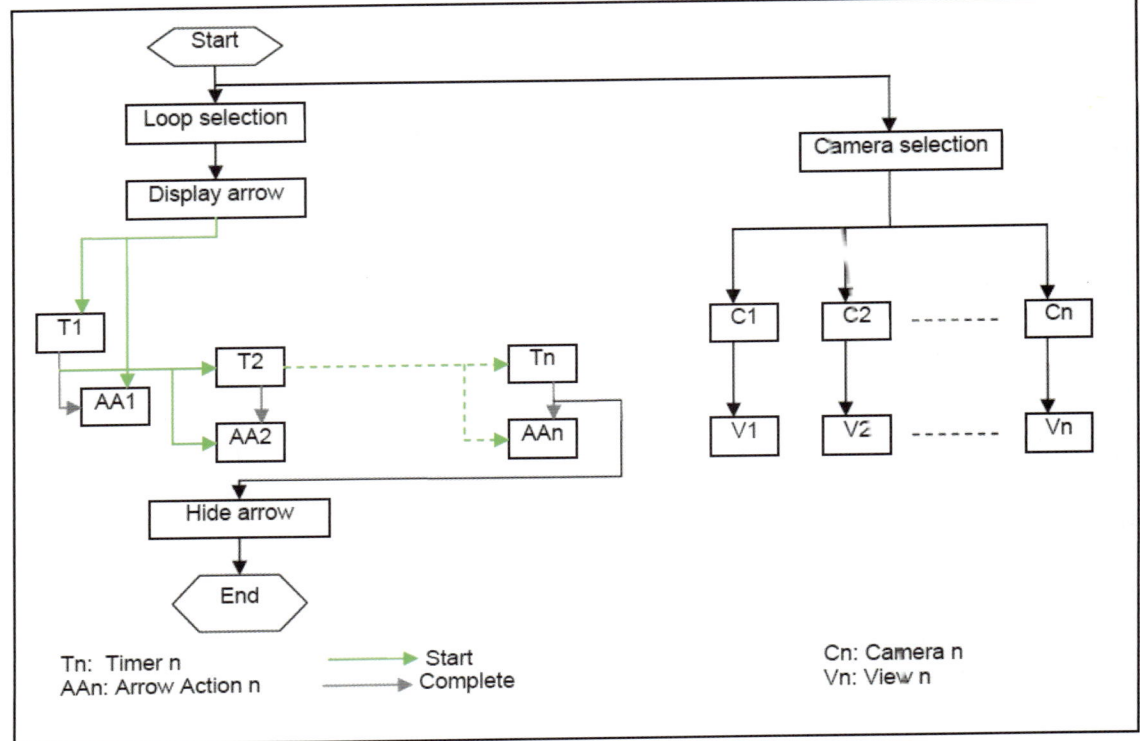

Figure 10. The procedure of the flow simulation.

product and then reassembled in the VR tools. The data extraction from CAD models to VR systems becomes a critical process of virtual product maintenance systems.

Product maintenance needs the product details, including the product bill of materials. It should support the analysis of "Design for X," such as design for manufacturing, design for assembly, and design for disassembly. An ideal product maintenance system should provide users not only with details of product design, manufacturing, and operations but also with product visualization, interactive processing, and information sharing. Therefore, the authors have proposed an integration method for product CAD models and VR-based systems in a Web-based environment to support product navigation, interactive maintenance, and data sharing.

A CAD-VRML (virtual reality modeling language) VR interface is developed for the visualization of product maintenance. The extraction of 3-D product information is based on CAD-VRML files, and the process incorporates the extracted data into a 3-D visualization environment. A MySQL database is used to store the extracted

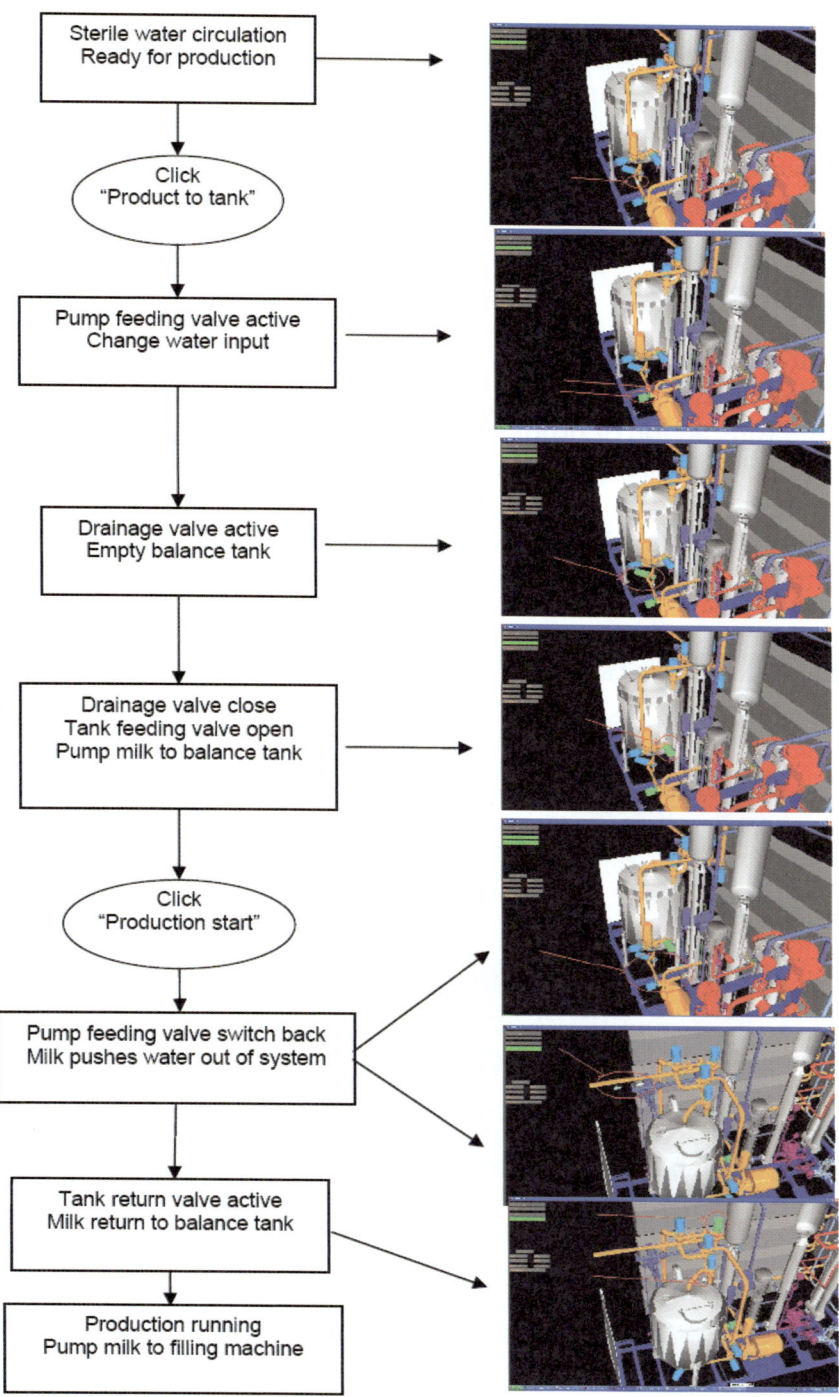

Figure 11. A simulation process of the production phase.

product information. A general translator of CAD-VRML data is implemented using Java techniques. The product maintenance operation is simulated by manipulating the geometry using a VRML external authoring interface (EAI). The interface has been used in product assembly and disassembly simulation and fixture planning (Kang & Peng, 2008). Figure 12 shows a simulation snapshot of a product disassembly operation using the proposed interface.

7   Conclusions

The high reliability of product maintenance systems is a result of good design, correct operation control, and monitoring. The increase of computer capabilities allows the visualization and interaction of product maintenance in a computer-created environment. This capability provides users an efficient tool to increase their productivity in product maintenance.

   This chapter presents a VR-based system to aid product maintenance training by enhancing user interaction. A feasible approach to enforce training of product

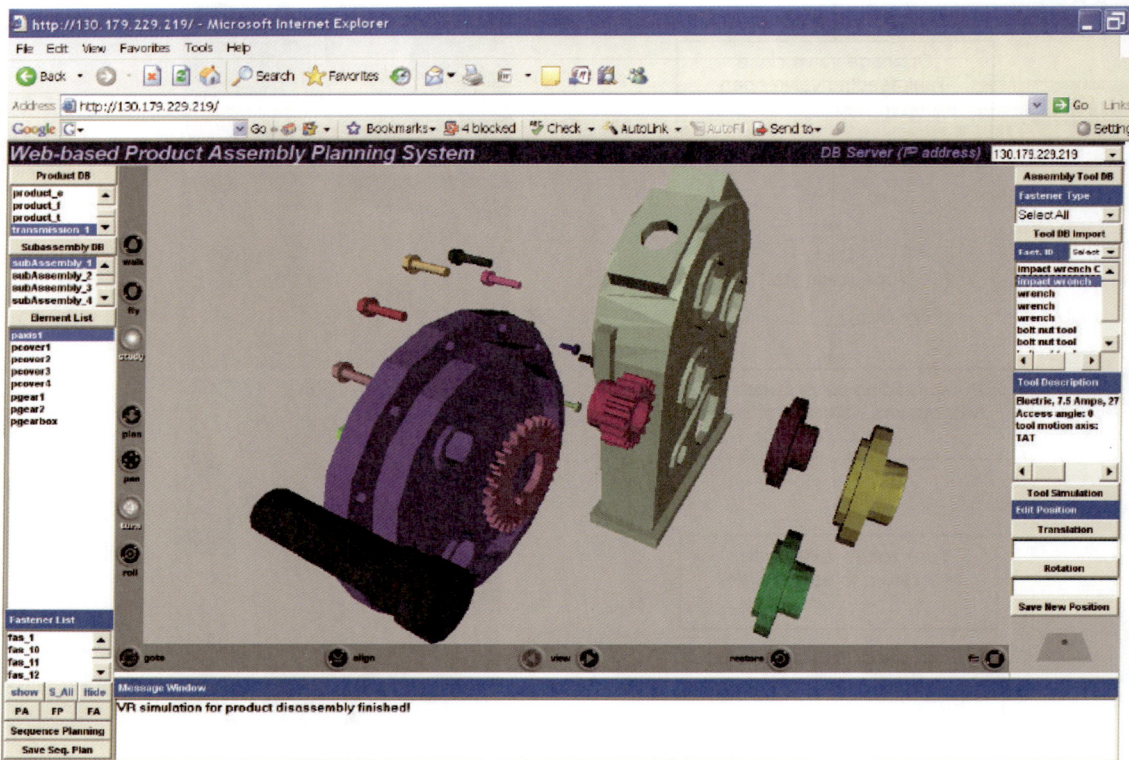

Figure 12. The simulation of a product disassembly sequence.

maintenance is discussed. The operation views of product maintenance are outlined in the examples. The emphasis of views is on the interdependencies of the product and components as well as the association with users. It improves current maintainability methods that do not support concurrent engineering. It also provides a direction for the integration of product design and maintenance analysis.

Acknowledgment

This research is supported by National Science and Engineering Research Canada (NSERC) Discovery Grants.

References

Cecil, J., & Kanchanapiboon, A. (2007). Virtual engineering approaches in product and process design. *International Journal of Advanced Manufacturing Technology, 31*(9–10), 846–856.

Chung, C., & Peng, Q. (2008). Enabled dynamic tasks planning in Web-based virtual manufacturing environments. *Journal of Computers in Industry, 59*, 82–95.

De Lit, P., Delchambre, A., & Henrioud, J. M. (2003). An integrated approach for product family and assembly system design. *IEEE Transactions on Robotics and Automation, 19*(2), 324–334.

EON Realty Inc. 2008. Retrieved July 23, 2008, from http://www.eonreality.com

Johansson, P., & Ynnerman, A. (2008). Immersive visual interfaces—assessing usability by the effects of learning/results from an empirical study. *Journal of Computing and Information Science in Engineering, Transactions of the ASME, 4*, 124–131.

Jude, K., Vinesh, F., Raja, H., & Eyre, J. (2003). Immersive learning system for manufacturing industries. *Computers in Industry, 51*, 31–40.

Kang, X., & Peng, Q. (2008). Fixture assembly planning in a Web-based collaborative environment. *International Journal of Internet Manufacturing and Services, 1*(2), 176–193.

Li, J. R., Khoo, L. P., & Tor, S. B. (2003). Desktop virtual reality for maintenance training: An object oriented prototype system (V-REALISM). *Computers in Industry, 52*, 109–125.

Low, K., Ilie, A., Welch, G., & Lastra, A. (2003). Combining head-mounted and projector-based displays for surgical training. *Proceedings of the IEEE Virtual Reality*, 1087–8270.

Marcelino, L., Murray, N., & Fernando, T. (2003). A constraint manager to support virtual maintainability. *Computers & Graphics, 27*, 19–26.

Peng, Q., & Yu, C. (2007). Enhanced integrated manufacturing systems in an immersive virtual environment. *Proc. Instn. Mech. Engrs, Part B: Journal of Engineering Manufacture, 221*(3), 477–487.

Puig, A., Perez-Vidal, L., & Tost, D. (2003). 3D simulation of tool machining. *Computers & Graphics, 27*, 99–106.

Reuding, T., & Meil, P. (2004). Predictive value of assessing vehicle interior design ergonomics in a virtual environment. *Journal of Computing and Information Science in Engineering, Transactions of the ASME, 4*, 109–113.

Smith, R. P., & Heim, J. (1999). Virtual facility layout design: The value of an interactive three-dimensional representation. *International Journal of Production Research, 37*(17), 3941–3957.

Wang, Q., & Li, J. (2006). Interactive visualization of complex dynamic virtual environments for industrial assemblies. *Computers in Industry, 57*, 366–377.

Waurzyniak, P. (2006, October). Collaborating with PLM. *Manufacturing Engineering, 137*(4), AAC1.

Yang, U., Lee, G. A., Shin, S., Hwang, S., & Son, W. (2007). Virtual reality based paint spray training system. *Proceedings of the IEEE Virtual Reality*, 289–290.

4 Creating Crane-Based Functionality in Virtual Environments to Facilitate Heavy Machinery Assembly Studies

Tatsuki Mitsui

Komatsu Ltd., Tokyo, Japan

Sankar Jayaram, Ph.D., Uma Jayaram, Ph.D., Frank Taylor, and Okjoon Kim

Washington State University, Pullman, WA

Abstract

We present the design and implementation of a fully functional overhead crane in a virtual assembly environment. This feature is essential for realistic assembly simulations of large-sized assemblies that are assembled using cranes; simulation environments that only use hands and handheld tools for assembly are inadequate. We discuss the crane geometry, the motion and degrees of freedom of the crane components, user interaction with the crane, user interaction with the part, and physically based modeling that allows gravity, reaction to forces, and collision detection. The results of an experiment comparing the virtual and real crane are also presented. Using this functionality, a module for the assembly of large press machines has been developed for Komatsu Ltd., an industry partner of the Virtual Assembly Technology Consortium. This module has been integrated with the existing virtual assembly design environment (VADE).

Keywords: crane simulation, physically based modeling, virtual assembly

1   Introduction and Motivation

Effective assembly-planning strategies and considerations such as assembly-planning algorithms, design for assembly principles, and integrated product and process design are important for industry. In particular, for assemblies involving large-sized components, assembly issues significantly impact the final cost of the product. Rework of inadequate facilities and faulty assembly sequences are cumbersome and expensive. Virtual reality (VR) technology is now being used to create assembly simulation environments to help manufacturing companies develop effective assembly procedures and reduce the labor, time, and cost involved with physical layout iterations and assembly sequence trials. It is important that these environments have the required functionality to allow realistic simulations of these very large parts.

The Virtual Assembly Technology Consortium was created to bring together industry, university, and government agencies to promote research, tools, and standards related to the use of VR and simulation to plan, evaluate, and verify assembly processes for mechanical systems. As a part of the consortium activities, a VR-based assembly application called VADE (virtual assembly design environment) was deployed to allow engineers to plan, analyze, and evaluate the assembly of mechanical systems. The environment, though useful and effective, was initially limited to assembly simulations of small assemblies that could be put together by hand or by handheld tools.

One of the founding members of the consortium, Komatsu Ltd., Japan, manufactures very large press machines for major automobile industries in different parts of the world. The press machinery is manufactured and semiassembled in house at the Komatsu site. After testing and troubleshooting, the equipment is then disassembled and shipped to the client site, where the heavy machinery is reassembled. This assembly process at the client site sometimes needs to be significantly modified on the fly due to layout differences between the client site and the Komatsu site. An enhanced virtual assembly environment that allowed simulation ahead of time of heavy machinery assembly using overhead cranes in an environment similar to the client's plant was essential to evaluate and troubleshoot the issues involved in reassembly of the large press machines. This chapter presents the crane-based functionality that was integrated with the virtual assembly environment to provide support for simulations involving heavy machinery in such situations.

2   Related Work

There is great interest in evaluating the practical significance of using VR technology for assembly planning and virtual prototyping (Choi, Chan, & Yuen, 2002; Yao, Ning, & Wang, 2002). Virtual assembly simulations provide a viable and productive tool (Chandrana, 1997; Gomes de Sá & Zachmann, 1999; Kibira & McLean, 2002). Some well-known virtual assembly applications are VADE (Jayaram et al., 1999; Jayaram, Jayaram, DeChenne, & Kim, 2004; Jayaram et al., 2007; Taylor, Jayaram, & Jayaram, 2006), developed at Washington State University; Virtual Design and Assembly System (VDAS; Fi, Choi, & Tu, 2002), developed at Hong Kong Polytechnic University; System for Haptic Assembly and Realistic Prototyping (SHARP; Seth, Su, & Vance, 2008), developed by Iowa State University; Virtual Training Studio (VTS; Brough et al., 2007), developed at University of Maryland; Haptic Assembly, Machining, and Manufacturing System (HAMMS; Lim et al., 2006), developed by Heriot-Watt University; and Collaborative Framework for Micro Assembly (Narayanasamy, Cecil, & Son, 2006), developed by New Mexico State University.

There are several applications that allow interaction with overhead cranes for simulation purposes. A PC application has been developed that allows the manipulation of a virtual crane and the simulation of crane motions (Wilson, Mourant, Man, & Weidong, 1998). Hoisting and conveying of machinery using VRML models have been supported (Wei, Tang, Rao, & Chen, 2007). A 3-D visualization capability has been integrated with a special-purpose simulation of a tower crane operation in a 3-D Studio MAX environment (Al-Hussein, Athar Niaz, Yu, & Kim, 2006). A virtual mock-up of a nuclear power plant has been developed, and a virtual crane allows simple movement of large objects (Whisker et al., 2002). A new training simulator of a container crane for port machinery that used VR and program control logic has been presented (Wang, Liang, & Liang, 2002). Commercial crane simulators are also available in the market (http://www.amc.csc.com/products/products _virtualcrane.htm; http://www.cranesafe.com/IES/nacb-ies-bridge-crane-systems .htm). Some researchers have exploited these virtual crane simulators to train operators in real time and have developed various interfaces (Rouvinen, Lehtinen, & Korkealaakso, 2005; Yoneda, Arai, Fukuda, Miyata, & Naito, 1997). These applications of a virtual crane range from large components in manufacturing industries to port machinery and log forwarders in forests.

3   Functionality Required for Simulation of Large Assemblies

As described before, the VADE application developed at Washington State University was used as the base immersive application to incorporate the proposed crane functionality. The first step was to understand and list the desired functionalities to be implemented for the proposed simulation. These are as follows:

- A realistic model of an overhead crane that can be imported into an immersive application with other parts such as jigs, fixtures, and hoists

- Methods to ensure realistic motion and degrees of freedom (DOF) of the crane

- Methods for the user to operate crane functions, control motion, and speed

- Methods to allow interaction with parts and reactions with external forces such as pushing with the hand or collision with other objects

- Physics-based modeling engine to simulate the pendulum motion of parts hanging from the crane and to check collision and space limitations in the environment

- A navigation system for the user to move around within the huge plant, since working space for the immersed user is restricted due to limitation of tracking systems

Figure 1 illustrates the different functionality and parameters involved in the simulation. The green boxes indicate the functionality desired, and other boxes indicate the parameters involved.

4   Design and Implementation of the Virtual Crane

The VADE developed at Washington State University was used as a platform to implement the crane functionality for assembly of heavy components. The problem was divided into the following subtasks. The first four will be discussed in this section, and the fifth one will be discussed in the next section:

1. Crane geometry
2. Motion/DOF of crane
3. User interaction with crane
4. User interaction with part
5. Physically based modeling

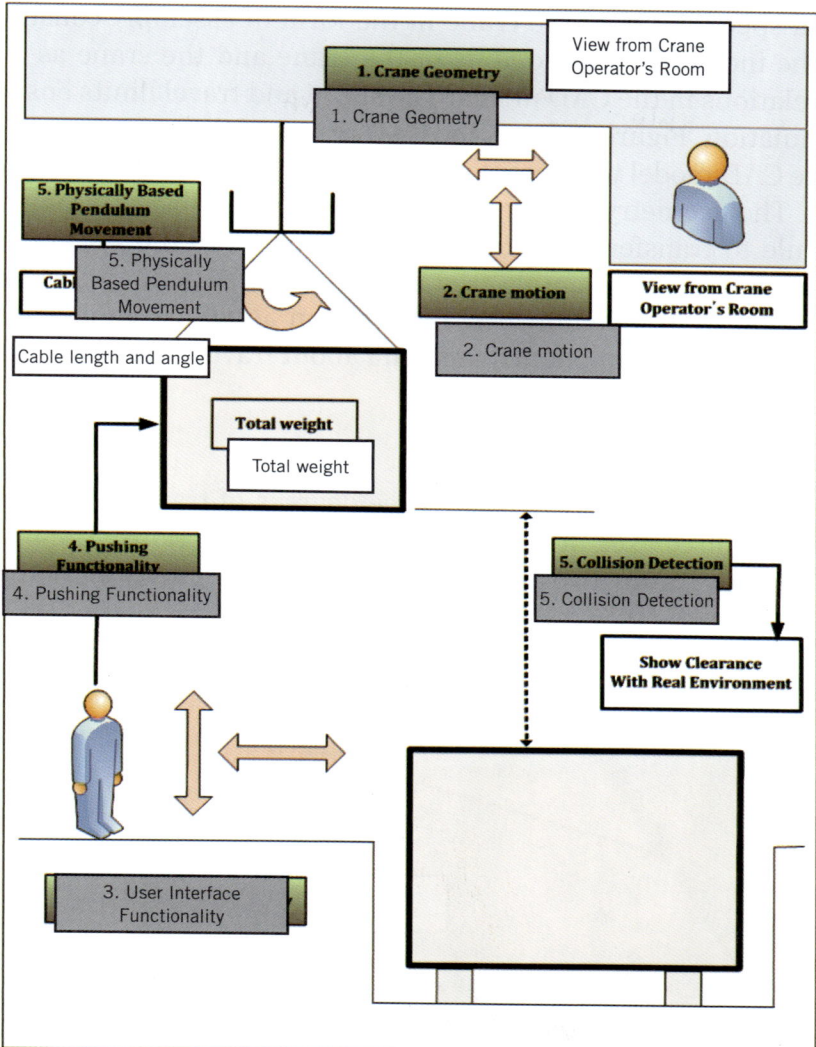

Figure 1. Proposed functionality.

4.1 Crane Geometry

The crane was first modeled in a computer-aided design (CAD) system by incorporating key pieces of crane geometry (Greiner, 1995), such as rails, the trolley, the bridge, the block, and the hooks. During the process of putting together the assembly, it was important to maintain a hierarchical representation of the crane by using parts and subassemblies appropriately so that the relative motions of the various portions of the crane would be correct. In addition, the CAD assembly also had information about the travel and speed limits of the crane in the form of assembly constraints. Limits of travel of the individual components of the crane and the crane as a whole were specified as relations in the CAD model. The speed and travel limits ensure a realistic assembly simulation. Figure 2 shows the CAD model of the crane geometry.

Once the CAD model was complete, this information was exported to the virtual environment. The geometry was exported into the immersive environment using an existing module to transfer data from the CAD system to the VR application. This is done by traversing the assembly tree inside a CAD system and exporting geometry files in the form of inventor files, information about assembly constraints and location/orientation of components, and data about travel and speed limits.

4.2 Motion/DOF of the Crane

For most of the components, the basic three degrees of translation have been considered along the three axes in positive and negative directions. The bridge and the trolley can each move along two axes. The hierarchical assembly of the crane allows a

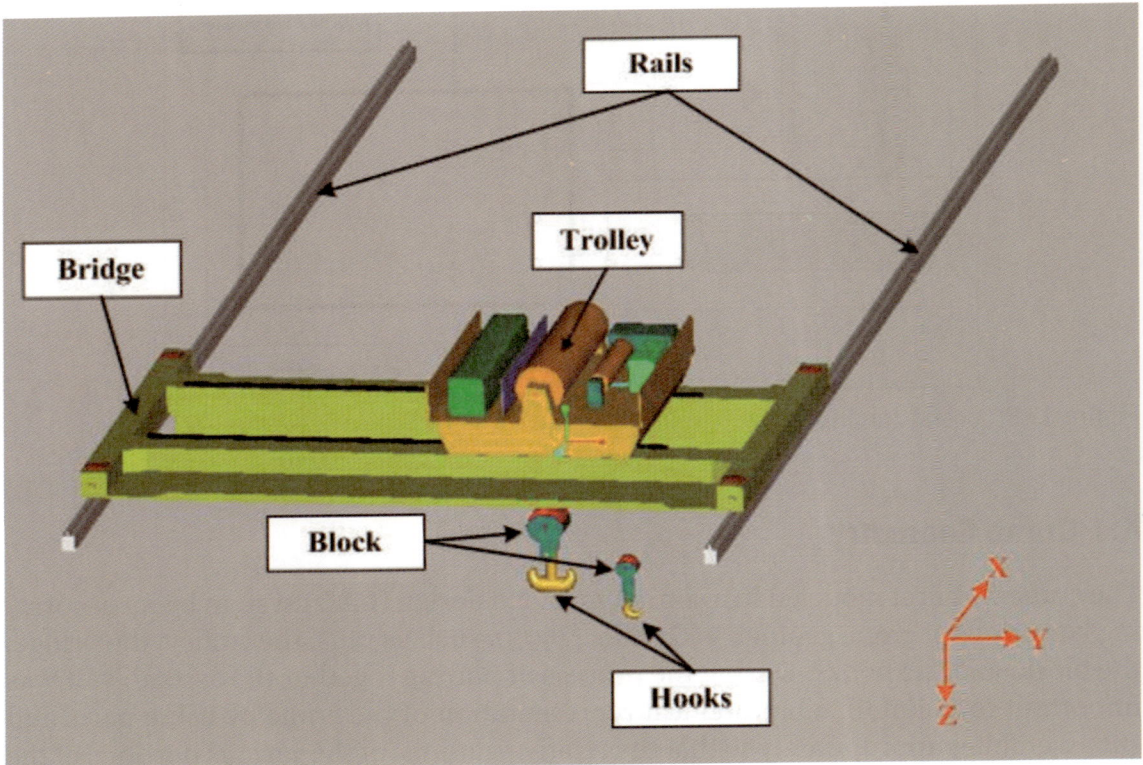

Figure 2. CAD model of the overhead crane geometry.

realistic motion simulation. For example, when the trolley of the crane moves along the rails of the bridge, the child components of the trolley in the assembly hierarchy, which are the blocks and hooks, will now move along with it. However, the bridge and rails components, which are the parents of the trolley in the hierarchy, do not move during the simulation. The "up" and "down" motion is applicable only to the hook. In addition, the hook can also rotate about its own axis. All the movements have a specified slow and fast velocity. The crane can be commanded by the movement of multiple directions at one time (e.g., north-east-up).

All of the local coordinate systems of the crane members are oriented so that the x-axis is parallel to the runway rails, the y-axis is parallel to the bridge rails, and the z-axis is pointed directly down. This simplifies calculations since the position of a crane member with respect to its parent only needs to be incremented or decremented along one axis when it is moved. For example, when the blocks and hooks move up or down, only the z-directional position value will be increased or decreased, respectively, with respect to the trolley.

4.3 User Interaction With the Crane

The user needs an interface to work with the crane in the immersive environment. The following key functionality was needed: (a) ability to control movement of the crane, (b) ability to control speed of the crane, (c) ability for the virtual human to move in the environment to reach the part that needs to be attached to the crane, (d) ability to specify the attachment points on the part that is to be attached to the crane hook, and (e) ability to attach the crane to the part. Different options were considered.

Our first idea was to create a virtual button box or pendant that would exist only in virtual space (graphical user interface [GUI] menu). In appearance, it would resemble a radio pendant. It would function similar to the virtual menus in VADE (Jayaram, Jayaram, & Mathrubutham, 1998) in that it would register a button-push event by detecting collisions between the geometry of the virtual hand and the geometry of the virtual pendant buttons. This was rejected for the following reasons. First, collision detection is a computationally expensive procedure. Second, the virtual menus at this time only register an instantaneous contact, and sometimes they do not even register this on the first try. The buttons on a real crane control pendant are usually pressed and held down for a period of several seconds. The inability of the virtual menu buttons to accurately register continuous contacts for a prolonged period of time would make it very difficult for this virtual button box to simulate an actual crane control pendant. Third, there is a possibility of accidentally triggering the wrong event during the performance of normal assembly tasks. Another option was to have a voice command system that would allow the user to speak commands like "attach" or "OK" that would trigger the events. This would give the benefit of allowing the user to directly give the commands in a "hands-free" manner. However, this required a reliable voice recognition system, and a few tests showed that this system was not very stable and efficient with the hardware and software available.

For our application, we decided that a dual-interaction method would work the best. We first created a physical control box, called the "button box" to resemble a real control box used on assembly lines. The button box is attached to the user around the waist by a belt. On its top face, it has an array of eight push-button switches and two toggle switches, which are easily accessible to the user. The eight push-button switches are used to control the direction of the crane's movement: four buttons for plant directions (north, south, east, and west) and two pairs of buttons to control up and down movements of the hook. The crane moves while a button is depressed and stops when the button is released. One toggle switch is used to indicate fast or slow speed. Figure 3 shows the layout of the button box. This functionality of the button box allowed the location and speed of the crane components to be controlled.

The button box is also used to maneuver through the virtual environment (i.e., a navigation tool for the immersed user). With the flip of a second toggle switch on the button box, the button box switches from control of the crane's location to control of the virtual human's location. This is beneficial for navigation in a large environment since the current human tracking system has a limited range of about 6 feet. With the button box, functionality is added to the virtual environment that lets the user's eye

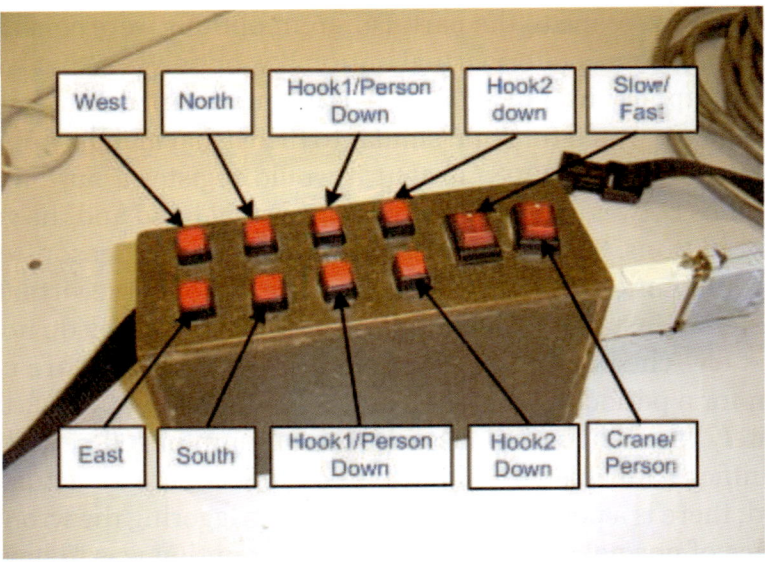

Figure 3. Button box.

point and hand location move to any location within the environment. Offsets would simply be added or subtracted to the translation coordinates supplied by the tracking devices. This functionality would give the user the ability to be accurately tracked at any position in the virtual environment.

In addition to the button box, there is an auxiliary keyboard command input system. For functionality (d) and (e) related to specifying the position of the attachment points and actually attaching the crane hook to the part, the VADE user verbally communicates with an assistant at the keyboard who then presses the appropriate key on the keyboard to select the command. This combination of the button box and the keyboard command input system worked well to fulfill all the requirements of user interaction.

4.4 User Interaction With the Part After It Is Attached to the Crane

As in the real environment, the user in the immersive environment has to interact with the part even after it is attached to the crane. This primarily involves pushing the part as needed while it is suspended from the hook. The pushing action could be used for the following purposes:

- To direct the part to a better position while it is suspended by the crane

- To rotate the part to guide it in situations involving space constraints

- To align the part in the final position (in the assembly)

The maximum force a person can exert needs to be simulated to prevent heavy parts from being easily rotated. Since this functionality is based on physically based simulation of the crane and the part, it will be discussed in more detail in Section 5.1.

5   Physically Based Crane Simulation

To properly evaluate an assembly process, real-time physics-based model simulation is necessary. Jansson, Vergeest, Kuczogi, and Kuczogi's (2001) approach used the mass-spring model representation extended with the concept of volume to generate a common algorithm for rigid-deformable or rigid-rigid body interaction. Wu (2001) talked about the method of mixed control of position/force in the assembly. This is a hybrid method suggested for effective physically based modeling in virtual assemblies. It is also important to see from a specified viewpoint if there is sufficient space available for an assembly in a narrow space. There is often a compromise between extra functionality and computational overhead.

For our simulation, it was important to model the behavior of objects being manipulated by the crane. This includes calculation of cable lengths, picking up a part using the crane, swinging while being moved around by the crane, reacting to contact with other parts, responding to a user pushing the component while it is hanging from the crane hook, and so on. The significance of this functionality is in having a realistic simulation of the process. The worker would want to know how a suspended part would behave when it is pushed to adjust its orientation or when it collides with the wall. A worker would also like to know whether there is enough space to rotate or tilt the part when it is assembled in a narrow area or if the cables are interfering with the assembly process. To determine such problems, VR assembly can be used effectively in conjunction with physically based equations.

The first step would be to identify the input and output parameters. The data extracted from CAD would be the inertia of the part and the center of mass. Other parameters extracted from the VR environment include the position of the crane hook, the position of hand, and so on. The output would be the suspended part's orientation and location.

The specification of attachment points on the part is an important step in preparing the part to be attached to the crane hook. The number of attachment points determines the pendulum motion of the part. If the part had one attachment point, it would act as a double pendulum swinging about the attachment point and the hook point. If it had three to four attachment points, the part would act as a single pendulum.

5.1 Modeling the Double-Pendulum Movement of the Crane

The physics of the movement of the crane hook was modeled as a double pendulum. Two data sets of hook locations are stored to calculate velocities for a small, defined time interval. From the change between two velocities, the acceleration of the hook is calculated:

$$\text{Acc} = \frac{\dfrac{[hp - hp_1]}{dt} - \dfrac{[hp_1 - hp_0]}{dt}}{dt},$$

where hp = the current hook positions, hp_1 = previous hook positions, hp_0 = previous to previous hook positions, and dt = time interval between frames.

The calculated data are then used for getting the orientation of the pendulum. Once the calculations are implemented and the frame is redrawn, the hook positions are updated to store new positions.

There are three Euler's angles to calculate the orientation of the pendulum. The initial position of the pendulum is shown in Figure 4. At first, the θ angle is defined and its value never changes during the simulation. This angle is to take into account

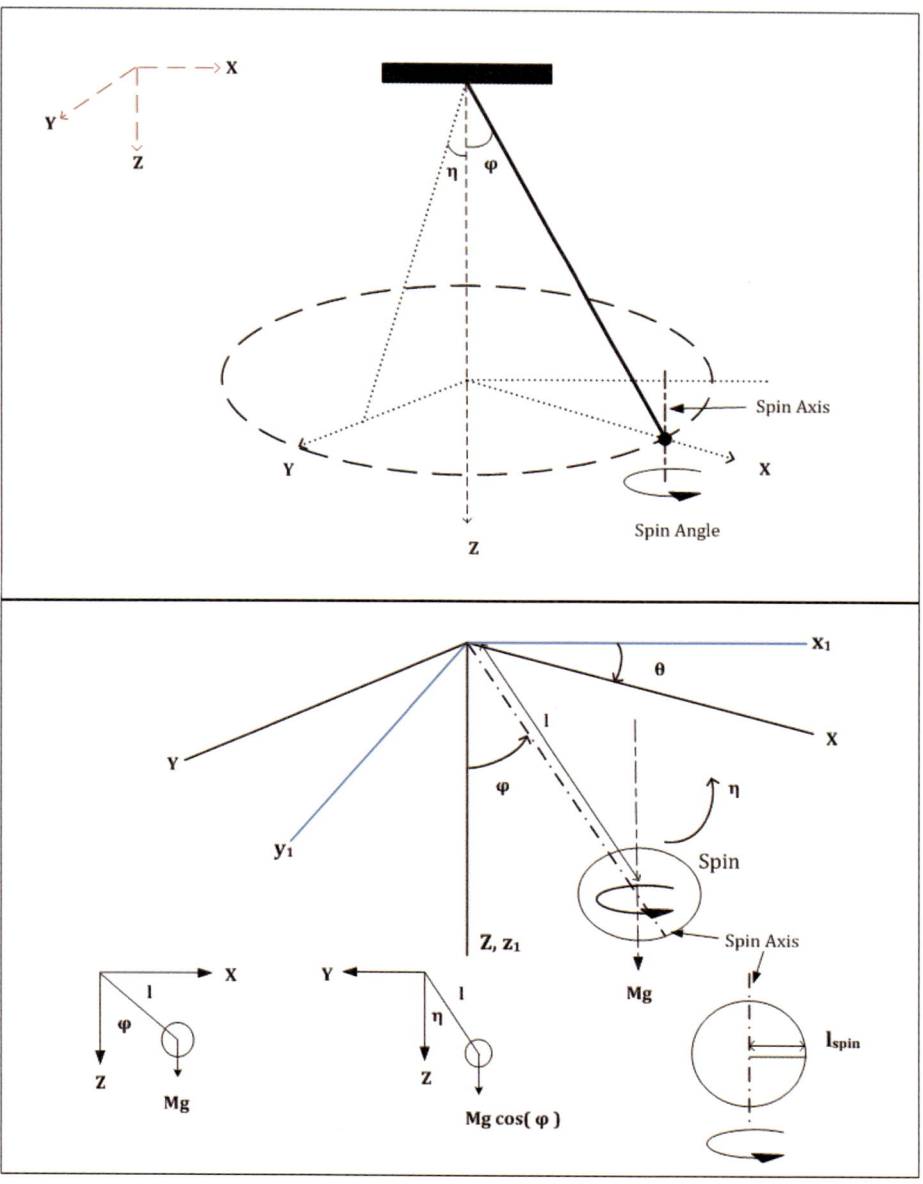

Figure 4. Specification of angles of the pendulum.

the mapping of the different coordinate systems—in our case, the hook and the component coordinate system. Next, the φ angle is defined, and its values are the initial values of the pendulum simulation. This movement is a plane movement with respect to the x-z plane, and its value will change according to the external forces such as gravity. The movement perpendicular to this movement is also defined, and in this case, η is an angle of this movement. An initial value is always 0 and changes according to the external forces. This movement is with respect to the y-z plane movement. The last angle that defines the pendulum movement is the spin angle. This angle will change when the suspended part is pushed by hand or hit by the wall. In such a case, the part would spin with respect to the axis, which goes through the pivot point to the center of mass. A final orientation of the pendulum is defined after the calculation of each of the following three angles.

The φ angle is defined by these equations:

$$\frac{d^2\varphi}{dt^2} + r\frac{d\varphi}{dt} + \frac{(M \times g \times l)}{(l_y)}\sin\varphi = -\frac{M \times l}{I_y}\left(\frac{d^2\alpha}{dt^2}\cos\varphi - \frac{d^2\beta}{dt^2}\sin\varphi\right) \tag{1}$$

$$I_y = I_{y(Globa\_Initial)} + (M \times I^2) \tag{2}$$

$$\frac{d^2\alpha}{dt^2} = \frac{d^2x_1}{dt^2}\cos\theta + \frac{d^2y_1}{dt^2}\sin\theta \tag{3}$$

$$\frac{d^2\beta}{dt^2} = \frac{d^2z_1}{dt^2} \tag{4}$$

The η angle is defined by these equations:

$$\frac{d^2\eta}{dt^2} + r_1\frac{d\eta}{dt} + \frac{(M \times g \times l \times \cos\varphi)}{(I_X) \times \sin\eta} = -\frac{(M \times l)}{I_X}\cos\varphi\left(\frac{d^2\alpha}{dt^2}\cos n - \frac{d^2\beta}{dt^2}\sin\eta\right) \tag{5}$$

$$I_x = I_{x (Globa\_Initial)} + (M \times I^2) \tag{6}$$

$$\frac{d^2\alpha}{dt^2} = \frac{d^2x_1}{dt^2}\sin\theta - \frac{d^2y_1}{dt^2}\cos\theta \tag{7}$$

$$\frac{d^2\beta}{dt^2} = \frac{d^2z_1}{dt^2} \tag{8}$$

Spin angle is defined by this equation:

$$\frac{d^2 spin}{dt^2} + r_2 \frac{dspin}{dt} = \frac{a_{spin} \times M \times l_{spin}}{I_{spin}}$$

(9)

The variables used are as follows:

M: Total weight of the hanged part

I_x: Inertia with respect to the *x*-axis

I_y: Inertia with respect to the *y*-axis

l: Distance between the pivot point and center of mass

I_{spin}: Inertia with respect to the axis that goes through the point of pivot point to center of mass

l_{spin}: Distance between the center of mass and the touched point

g: Gravity

spin: Angle of the spin

a_{spin}: The spin acceleration from the hand

r, r1, r2: Damping coefficient of the swing

These equations assume that there is no second order of inertia in the suspended part. Only first-order inertia is counted even if there is a second order of inertia.

To get the orientation, the fourth-order Runge-Kutta method is used for calculation. After getting the angles, the pendulum location is calculated by using these values. These values are used to update the frame in the virtual environment to get the new orientation of the hook and the part.

5.2 Modeling the Pushing Functionality

In order to model the pushing functionality that allows the user to change the orientation of parts while they are suspended, different criteria were defined, as shown in Table 1. The criteria are based on different factors, such as the relative movement of the part and the user, the weight of the part, and so on. Accordingly, a pushing event was generated.

Table 1. Criteria of Pushing Force

No.	Event	Action
1	Part moves freely or part is close to stopping near the stable position but is still moving	No force is applied
2	Part is moving away while hand is pushing	Pushing force is applied
3	Part is moving away while hand is pushing back	No force is applied
4	Part is moving toward the operator while hand is pushing	If weight <100 kg, hand can stop the part; if weight >100 kg, hard cannot stop the part
5	Part is moving toward the operator while the hand is pushing back	If weight <100 kg; hand can stop the part; if weight >100 kg, hand cannot stop the part

The pushing point is calculated by the average value of the touching hand sensor data. The force from a pushing hand is derived from the acceleration of the hand and some appropriate weight. The force has two components. One changes the angles φ and η, and the other changes the spin angle.

Acceleration of the hand is calculated by using location and time intervals (Figure 5). This acceleration is divided into two components. One component is along the vector from the CG (center of gravity) to the touching point, and the

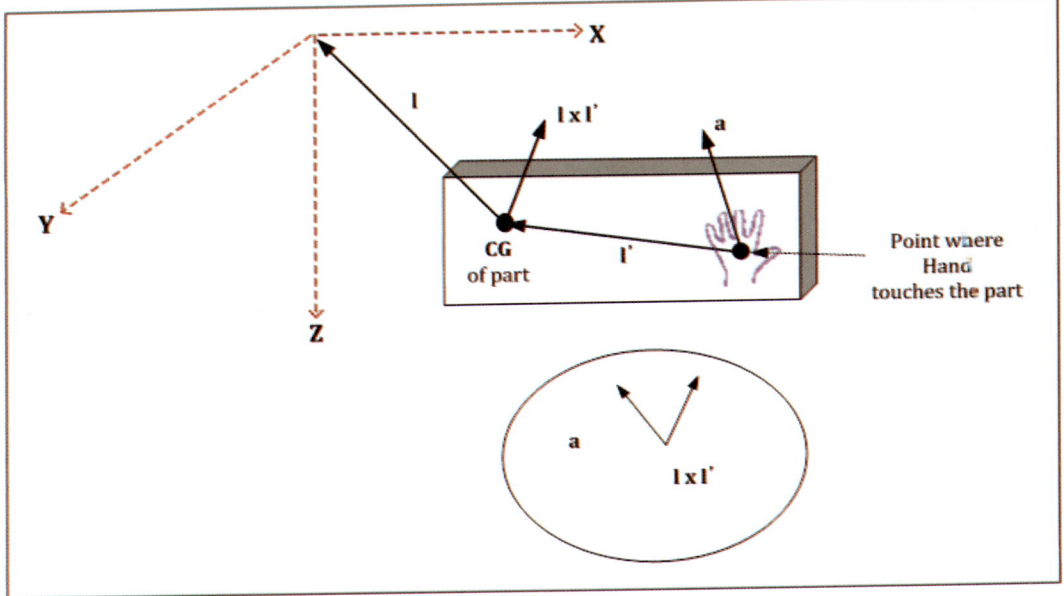

Figure 5. Direction of the application of the pushing force.

other is derived by subtracting this component from the original force vector. $\vec{\imath}, \vec{\imath}\,'$, and $\vec{a}$ are vectors. The angle between $(\vec{\imath} \times \vec{\imath}\,')$ and $\vec{a}$ is defined as

$$angle = cos^{-1}((\vec{\imath} \times \vec{\imath}\,') \bullet \vec{a}),$$

and acceleration of the original $\vec{a}$ is divided into two components:

$$\vec{a}_{spin} = |\vec{a}|\, cos(angle)\, (\vec{\imath} \times \vec{\imath}\,')$$

$$\overrightarrow{a_{push}} = \vec{a} - \overrightarrow{a_{spin}}$$

($\overrightarrow{a_{spin}}$ and $\overrightarrow{a_{push}}$ are vectors).

The acceleration points at CG are divided for acceleration for movement about ϕ and η. After these values are calculated, all that is required is to multiply the mass to obtain the force. If the mass of the part is less than 100 kg, the mass is multiplied to obtain the defining force. If the mass of part is over 100 kg, the mass multiplied for defining force is set to 30 kg.

The absolute velocity of the hand itself is also calculated in order to prevent an unintentional call to the pushing event. Without calculating the absolute velocity of the hand, a force is applied if the human is moving, even if the hand is not moving. The problem is that the force is calculated by the touching point between hand and part. It is possible for the hand's palm to touch the part and one of the fingers to touch a different point of the part at next frame. This produces an inappropriate force violating the assumption that the touching point does not change quickly. Ideally, the hand should push the same point with respect to the local coordinate systems of the part. This case happens sometimes when the user tries to stop the swinging part. The user will see that the part's swing is going to increase gradually instead of going to stop. To solve this problem, the algorithm is added to ensure that the force is not going to be applied to the part if the hand is not moving.

6 Implementation

The physics-based crane functionality was implemented in VADE. Models provided by Komatsu were used for the simulation. The CAD models with mass properties were exported to VADE. Figure 6 shows the new realistic environment as it appears in VADE. This environment resembles a plant from Komatsu Ltd. It was necessary to design a method for integrating the virtual crane into the VADE application. This

Figure 6. Crane implementation in immersive environment (VADE).

integration took place in two major steps. The first step was to make the virtual crane visible and controllable from within the VADE application. The second step of integration was the task of implementing the complex interactions between the Crane class and the Part class in VADE.

New classes were added to integrate the crane functionality into VADE. This section describes the integration of the Crane class and the Pendulum class with the Part classes. The Part class represents the parts in the assembly model and contains all the constraint information and dynamic state information, and the Pendulum class is used for the simulation of the pendulum motion of a part attached to the crane. To have a successful integration of the Crane, Part, and Pendulum classes, several interactions have to be considered:

- Providing the location of the Crane hook point and Part center of mass to the Pendulum
- Displaying the Part motion based on the Pendulum output

- Specifying Part attachment points

- Attaching and detaching the Part to and from the Crane

- Displaying cables attaching the Part to the Crane hook

- Displaying Part weight and cable tensions

One of the major challenges of these interactions was the difference in coordinate systems between the crane, part, and pendulum.

In order to calculate the position of a part attached to the crane, the Pendulum class has to be updated at each frame with the locations of the crane hook point and the center of mass of the part. It required coordinate system transformations to bring the crane hook point and the part center of mass into the same coordinate system, since the two were at different points in the scene graph. It seemed that there were two solutions:

1. When attaching a part to the crane hook, apply the inverse transform of the crane hook to the part and make the part a child of the crane hook. This way, the two points are in the crane hook coordinate system.
2. Apply the transformations of the crane hook and all its parents to the location of the hook point. This puts both points in global coordinates.

We decided to go with the latter choice, since it is less complicated.

The purpose of the Pendulum class was the simulation of the pendulum motion of a part attached to the crane. The pendulum returned as output the four rotation angles (x-y-z and spin) and the x-y-z position used to determine the orientation and position of the part.

The following list shows the order of transformations applied to the attached part to properly position it:

- Translate the part's center of mass to the origin.

- Give the part the orientation it had when it was first attached to the crane.

- Apply the VADE-to-Pendulum transform (rotate 180 degrees about x).

- Apply the initial theta rotation about the negative z-axis. (Theta rotation refers to the angle between the x-axis and the line cast by the pendulum arm on the x-y plane when the part is first attached to the crane. The Pendulum class treats the line cast on the x-y plane as its primary axis of rotation and the line perpendicular to it as the perpendicular axis of rotation.)

- Apply the x-y-z rotations supplied by the pendulum.

- Apply negative initial theta rotation about the negative z-axis.

- Apply spin to the part about the axis going from the hook attachment point to the part center of mass.

- Translate the part to the pendulum-specified part location.

- Apply the Pendulum-to-VADE transformation.

For added realism, we added the display of cables going from each of the part's attachment points to the crane hook point. In the scene graph hierarchy, the cables, as the attachment points, are children of the part. Examples of the attachment points and cables for the single and double pendulums can be seen in Figures 7 and 8.

7   Test Case Results

A simple comparative study was carried to compare the simulated motions of the crane and attached component in the virtual environment with motions of a real crane and attached component. Detailed information of the experiment can be found in Jayaram, Taylor, Jayaram, and Mitsui (2000). The newly created functionality of the heavy machinery module was submitted to two types of testing:

1. Performance of assembly tasks in the virtual environment that had already been performed in the real environment, verifying that virtual assembly tasks can be completed using the heavy machinery module
2. Comparison between the physical behaviors of the virtual crane and pendulum and an actual crane and pendulum

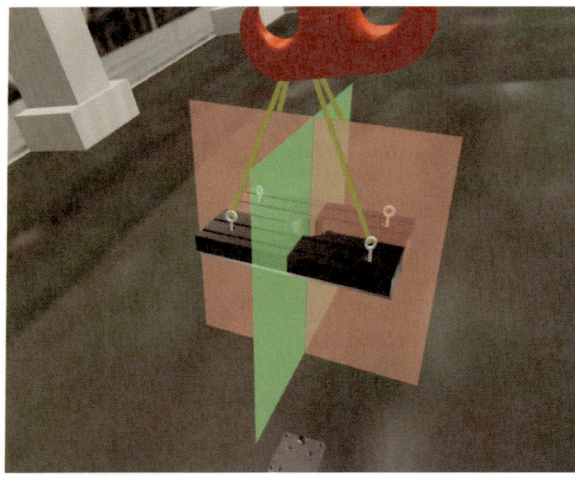

Figure 7. Single-pendulum motion.

Figure 8. Double-pendulum motion.

A comparison of the results of the physical and virtual experiments is shown in Table 2.

As Table 2 indicates, there are several obvious differences between the calculated values of the physical experiment and the virtual experiment. The following are possible explanations for those differences:

- *Acceleration.* This error arises from the assumption of constant acceleration. Because the measurement of time in the virtual environment is discrete, the acceleration in the time step just before maximum velocity is reached is less than the specified acceleration.

- *Damping factor.* One possible source of error is sample rate. Angular displacement values may not have been recorded at the exact apexes of the two measurements taken. This is compounded by the fact that there was relatively little difference between the angular displacement values read, since reading was taken for only 20 oscillations and the specified damping factor is quite small. Though

Table 2. Comparison of Results Between Physical and Virtual Experiments

	Physical	Virtual	Error
Max. velocity (in./sec.)	21.06	21.06	0%
Acceleration (in./sec.2)	146.28	165.23	13%
Damping factor	2.48E-04	6.34E-05	74%
Angular deflection from acceleration (deg.)	6.80	7.84	15%

there is an order of magnitude difference between the damping factors, both the physical and virtual pendulums move in an almost nondamped fashion.

■ *Angular displacement due to acceleration.* Errors between these values could arise from the accuracy of the device used to measure angular displacement in the physical experiment. The measurement device was accurate at best to one degree.

8 Conclusion

Assembly planning relies mostly on skilled workers' expert knowledge of assembly processes. Virtual assembly along with crane functionality can prove to be a useful tool for this purpose rather than physical mock-ups or hand drawings. The crane and the button box have been designed and implemented successfully in VADE. The test cases have shown that the newly developed technology can be used to virtually simulate large assemblies that are performed in the real world. Assembly of two press machines was carried out for Komatsu using the crane functionality in VADE (Jayaram et al., 2007; Jayaram et al., 2004; Taylor et al., 2006).

More work can be done to improve the level of realism. The following are some ideas:

■ Force feedback device can be used to give the user a sense of resistance for heavy parts

■ Complete control for the user in the immersive environment (now we use a combination of button box and keyboard commands)

■ Ability to switch the models in a different environment to evaluate space constraints for assembly in different plants

■ Implementation of a complete human model

■ Integration with other simulation toolkits to generate a "knowledge base"

■ Effective and computationally less expensive collision detection to avoid a situation like a hand penetrating a part

With all these add-ins, virtual assembly for large, heavy machinery will be more realistic and hence more reliable and beneficial to the industry.

Acknowledgments

This work has been funded by the Virtual Assembly Technology Consortium at Washington State University and its founding members—Caterpillar Inc., Deere & Company, Komatsu Ltd., PACCAR Inc., and NIST.

References

Al-Hussein, M., Athar Niaz, M., Yu, H., & Kim, H. (2006). Integrating 3D visualization and simulation for tower crane operations on construction sites. *Automation in Construction, 15*(5), 554–562.

Brough, J. E., Schwartz, M., Gupta, S. K., Anand, D. K., Kavetsky, R., & Pettersen, R. (2007). Towards development of a virtual environment-based training system for mechanical assembly operations. *Virtual Reality, 11*(4), 189–206.

Chandrana, H. S. (1997). *Assembly path planning using virtual reality techniques* (Master's thesis). Washington State University, Pulman, WA.

Choi, A. C. K., Chan, D. S. K., & Yuen, A. M. F. (2002). Application of virtual assembly tools for improving product design. International *Journal of Advanced Manufacturing Technology, 19*(5), 377–383.

Fi, P., Choi, A. C. K., & Tu, L. (2002). VDAS: A virtual design and assembly system in a virtual reality environment. *Assembly Automation, Emerald, 22*(4), 337–342.

Gomes de Sá, A., & Zachmann, G. (1999). Virtual reality as a tool for verification of assembly and maintenance processes. *Computers & Graphics, 23,* 349–403.

Greiner, H. G. (1995). *Crane Handbook 2nd Edition.* Harvey, IL: Whiting Corporation.

Jansson, J., Vergeest, J. S. M., Kuczogi, G., & Horvath, I. (2001). Merging deformable and rigid body mechanics simulation. *Proceedings of the Fourteenth Conference on Computer Animation,* Seoul, South Korea, 147–56.

Jayaram, U., Jayaram, S., DeChenne, C., & Kim, Y. J. (2004, September/October). *Case Studies Using Immersive Virtual Assembly in Industry.* Proceeding of ASME 2004 DETC & CIE Conference, Salt Lake City, Utah.

Jayaram, S., Jayaram, U., Kim, Y. J., DeChenne, C., Lyons, K. W., & Mitsui, T. (2007). Industry case studied in the use of immersive assembly. *Virtual Reality, 11*(4), 217–228.

Jayaram, S., Jayaram, U., & Mathrubutham, N. (1998, September). *Creating and managing virtual menus in immersive environments.* ASME International Computers in Engineering Conference, Proceedings of DETC 98, Atlanta.

Jayaram, S., Jayaram, U., Wang, Y., Tirumali, H., Lyons, K., & Hart, P. (1999). VADE: A virtual assembly design environment. *IEEE Computer Graphics and Applications, Institute of Electrical and Electronics Engineers Computer Society, 19*(6), 44–50.

Jayaram, S., Taylor, F., Jayaram, U., & Mitsui, T. (2000, September). *Functionality to facilitate assembly of heavy machines in a virtual environment.* Proceedings of ASME DETC 2000, Baltimore.

Kibira, D., & McLean, C. (2002). Virtual reality simulation of a mechanical assembly production line. *Proceedings of the 2002 Winter Simulation Conference, 2,* 1130–1137.

Lim, T., Calis, M., Ritchie, J. M., Corney, J. R., Dewar, R. G., & Desmulliez, M. (2006, March). A Haptic Assembly, Machining and Manufacturing System (HAMMS) Approach, *1st International Virtual Manufacturing Workshop* (VirMan '06), Virginia.

Narayanasamy, G., Cecil, J., & Son, T. C. (2006). A collaborative framework to realize virtual enterprises using 3APL. *Declarative Agent Languages and Technologies IV, 4327*, 191–206.

Rouvinen, A., Lehtinen, T., & Korkealaakso, P. (2005). Container gantry crane simulator for operator training. *Journal of Multi-body Dynamics, 219*(4), 325–336.

Seth, A., Su, H., & Vance, J. M. (2008). Development of a dual-handed haptic assembly system: SHARP. *ASME Journal of Computing and Information Sciences in Engineering, 8*(4), 044502–1 to 044502–8.

Taylor, F., Jayaram, S., & Jayaram, U. (2006). Validation of virtual crane behavior through comparison with a real crane. *International Journal of Advanced Manufacturing Systems (IJAMS), 9*(1), 73–83.

Wang, C., Liang, G., & Liang, C. (2002). The training simulator of container crane. *Journal of System Simulation, 14*(7), 904–906, 921.

Wei, G., Tang, Q., Rao, G., & Chen, D. (2007, July). Research and implementation of the operatable virtual models of hoisting and conveying machinery. *International Conference on Transportation Engineering 2007*, Chengdu, China, *4*, 3170.

Whisker, V. E., Baratta, A. J., Mouli, S. C., Shaw, T. S., Warren, M. F., Winters, J. W., & Clelland, J. A. (2002). Modular Construction Installation Study Using Virtual Environments. *Transactions of the American Nuclear Society, 87*, 41.

Wilson, B. H., Mourant, R. R., Man, L., & Weidong, X. (1998). A virtual environment for training overhead crane operators: real-time implementation. *IIE Transactions, 30*(7), 589–595.

Wu, W. (2001). Virtual Assembly Model With Mixed Control of Position/Force. *Journal of Beijing University of Aeronautics and Astronautics, 27*(4), 377–380.

Yao, J., Ning, R., & Wang, X. (2002). Research of virtual reality based assembly planning. *Jixie Gongcheng Xuebao/Chinese Journal of Mechanical Engineering, Editorial Office of Chinese Journal of Mechanical Engineering, 38*(8), 130–134.

Yoneda, M., Arai, F., Fukuda, T., Miyata, K., & Naito, T. (1997, September/October). *Operational assistance system for crane using interactive adaption interface—design of 3D virtual crane simulator for operation training.* Robot and Human Communication, 1997 (RO-MAN '97). Proceedings of 6th IEEE International Workshop, 224–229.

An Efficient and Scalable Haptic Modeling Framework for Needle-Insertion Simulation in Percutaneous Therapies Training System

Jing Qin, Yim-Pan Chui, Wing-Yin Chan, and Pheng-Ann Heng, Ph.D.
Department of Computer Science and Engineering,
The Chinese University of Hong Kong

Simon C. H. Yu
Department of Diagnostic Radiology and Organ Imaging,
The Chinese University of Hong Kong

Abstract

An efficient and scalable framework for interactive haptic modeling and simulation of needle insertion is described. An integrated solution is implemented to model the prepuncture forces, friction forces, and cutting forces when the needle penetrates into skin, adipose tissues, muscle, and internal organs such as the liver. A client-server-based distributed visual-haptic framework is developed to parallelize visual and haptic rendering processes in order to avoid performance overhead occurring in concurrently running these processes on the same machine. The proposed framework has been adopted to develop two medical training systems: an ultrasound-guided organ biopsy simulation system and a Chinese acupuncture training system. The success of these applications demonstrates the feasibility of the proposed framework.

Keywords: surgical simulation, virtual medicine, visualization

1 Introduction

Many percutaneous therapies, such as biopsies, anesthesia drug injections, and Chinese acupuncture, require inserting a needle through skin to a specific target of internal soft tissue. It is a fundamental but difficult skill in modern clinical practice. Although real-time imaging techniques have been employed to guide the insertion procedure in some of these procedures, physicians and surgeons usually rely on kinesthetic feedback from the needle to judge if the operation is performed correctly, especially when the visual feedback cannot provide sufficient or exact information about the location of the needle. Considerable training and practice are required for a student or novice to acquire this skill.

Traditionally, the training of this skill has been performed on live patients, cadavers, and animals. However, practicing on patients not only jeopardizes their health but also provides limited access to training scenarios and makes it difficult for training in a time-efficient manner. On the other hand, cadavers lack tissue realism, and the limited availability of cadavers is another big concern. The use of animals cannot be considered as the perfect substitute for the human body due to anatomical differences between animals and humans as well as other ethical issues.

Recent years have witnessed significant progress in surgical simulation systems based on virtual reality (VR) for training novices and performing surgical planning and rehearsal for complex procedures. Besides realistic visualization of human anatomy, many recently developed simulators have integrated kinesthetic feedback through the use of haptic devices. A lot of research work has demonstrated that integrating visual and haptic sensation can greatly enhance the effectiveness of the VR-based surgical simulation system (Basdogan et al., 2004; Niemeyer et al., 2004).

Some simulators have been developed for the training of various percutaneous therapies. While visualization of human anatomy and deformation of soft tissue are realistically provided, most simulators lack an integrated modeling and simulation solution for interactive haptic rendering, especially when multiple tissues with heterogeneous biomechanical properties are involved. They either focus on only a specific soft tissue or are customized to a target application and thus cannot be easily reused in the development of other systems. In addition, most simulators can support only one haptic device. In reality, a lot of interventions are performed by two hands or even several surgeons. To obtain realistic and stable haptic sensation, the haptic rendering loop must maintain a 1000 Hz update rate, which is much higher than that of graphic rendering (about 20–30 Hz). In addition, the disparity in the update rate between the graphics loop and the haptics loop makes it easy to display an inconsistent state in the virtual environments. This requirement to display a consistent state makes haptic rendering a more demanding task. The situation is more stringent when multiple haptic devices are involved.

In this chapter, we report our experience in the development of an efficient and scalable framework for interactive haptic modeling and simulation of needle insertion and its application in training simulators for percutaneous therapies. An integrated solution is implemented to model the various force components when the needle penetrates skin, adipose tissues, muscle, and internal organs. Prepuncture forces, friction forces, and cutting forces are all realistically and interactively modeled in our framework. In addition, after puncturing the skin, path constraint force and torque are simulated by employing haptic devices with 6-degrees of freedom (DOF) force feedback to prevent the virtual needle from deviating from the original trajectory. A client-server–based distributed visual-haptic framework is developed to parallelize visual and haptic rendering in order to minimize performance overhead occurring in concurrently running haptic and visual rendering processes on the same machine. Our framework has been adopted to construct two medical training systems: an ultrasound-guided organ biopsy simulation system and a Chinese acupuncture training system. The success of these applications demonstrates the feasibility of our framework.

2 Related Work

Existing VR-based simulators have proposed various solutions to model and simulate the needle-insertion procedure in percutaneous therapies. Dang, Annaswamy, and Srinivasan (2001) proposed an epidural injection simulator with force feedback for medical training and education. However, it can only provide three-dimensional force feedback where the initial needle trajectory cannot be well maintained during the training procedures. Kwon et al. (2001) implemented a system to produce force reflection for needle insertion in a spine biopsy simulator. But it cannot properly simulate prepuncture forces and cutting forces, which are important force components during the insertion procedure. Gorman, Krummel, Webster, and Hutchens (2000) proposed a haptic lumbar puncture simulator, which provides tactile feedback for the training procedure. Unfortunately, it does not consider the biomechanical properties of different tissues and accordingly cannot provide realistic force sensations. Other simulators have been developed for breast biopsy (Azar, Metaxas, & Schnall, 2000), liver biopsy (Forest, Comas, Vaysière, Soler, & Marescaux, 2007; Magee, Zhu, Ratnalingam, Gardner, & Kessel, 2007), and prostate needle biopsy (Zeng et al., 1998). But most of them are limited to specific aspects or have the aforementioned shortcomings and cannot fulfill all requirements of needle-insertion procedures. In this regard, a systematic solution is needed (a) to realistically render multiple force components during needle-insertion procedures, (b) to consider multiple tissues with heterogeneous biomechanical properties, and (c) to support multiple haptic devices without degrading the system's performance.

3 Integrated Haptic Modeling

3.1 System Requirements

Several important factors were considered during the implementation of our framework:

- *Realism.* One of the goals in our framework is to construct models that can demonstrate realistic force behaviors during needle-insertion procedures. In this regard, we employed many experimental results and models in biomechanical literature as references.

- *Computation.* Computational complexity in simulation has to be considered. The computational requirement for haptic rendering is especially tight. To achieve a stable haptic display, an update rate of more than 1 kHz is desired. Besides, the purpose of our simulation is percutaneous therapies training; having an interactive update rate is the key to improving students' performance, unlike simulation for surgery planning and scientific analysis. Since the requirements of realism and computational efficiency always conflict with each other, a trade-off has to be made between the two in the simulation design. So, in designing our haptic models, we tend to seek simple equations that can describe the experimental results instead of computationally expensive models of complex tissue behaviors.

- *Multiple force components.* In order to accurately model the feedback forces, it is necessary to characterize the typical force profile of needle insertion from skin into internal organs. Generally, there are two or three force peaks in the force-displacement relationship during the procedure, indicating the puncture of skin, muscle, and the internal organs' capsule. In each peak, there is a rise of force resulting from the stiffness of tissues, followed by a sudden drop of force during puncture. After every puncture event, the force fluctuates with a slightly increasing trend due to the increasing friction resulting from the gradual increase in the contact area between the needles and underlying tissues. When the needle penetrates internal organs such as the liver, we need to consider forces caused by cutting internal structures such as arteries or veins. Based on this typical needle-force transience, we implement three models to simulate the prepuncture force, friction force, and cutting force, respectively. In addition, after puncturing the skin, path constraint force and torque are simulated to prevent the virtual needle from deviating from the original trajectory. Finally, we composite the total force and torque in every haptic frame by adding them all.

■ *Multiple haptic devices.* While many simulators have been developed with haptic sensation, most of them can support only one haptic device. In reality, a lot of interventions are performed by two hands or even several surgeons. For example, in ultrasound-guided biopsy, surgeons manipulate an ultrasound transducer and a biopsy needle collaboratively to insert the needle into an accurate position. The success of this procedure is dependent on the correct alignment of the biopsy needle with the scanning plane of the ultrasound transducer and with the target lesion. To simulate these scenarios, multiple haptic devices should be supported without degrading the system's performance in the virtual environment.

■ *Multiple tissues.* In general, multiple tissues with heterogeneous biomechanical properties are involved in percutaneous therapies, such as skin, adipose tissues, muscle, and internal organs. Different parameters in our haptic models have to be calibrated before starting the training procedure, such as the elastic and viscous coefficients of involved tissues and coefficients of kinetic friction of involved tissues.

3.2 Haptic Modeling

3.2.1 Prepuncture Force Modeling

We employed the incremental viscoelastic model developed by Brett, Parker, Harrison, Thomas, and Carr (1997) to model the prepuncture forces of skin, muscle, and the internal organs' capsule. It is evident that this model can well approximate the character of the prepuncture forces, and its computational cost is relatively low compared to some finite element models. The principle of the model is shown in Figure 1a. It starts with a single Voigt element; a new Voigt element is added each time the total needle displacement x exceeds a multiple of an adjustable variable, Δx. The force required to make a displacement of x before puncture is

$$F_{pp}(x,v) = \frac{x(6\nu v \Delta x + kx^2 + 3kx\Delta x + 2k\Delta x^2)}{6\Delta x^2},$$

(1)

where k and v are the elastic coefficient and viscous coefficient of the tissue, while v is the velocity of the needle. According to this equation, when the needle displacement x and the velocity of the needle movement v are greater, users will feel greater resistant forces from the haptic devices.

3.2.2 Friction Force Modeling

The friction forces occur along the needle-insertion path and are caused by the relative motion between the needle and tissues. Due to the needle clumping, there exists a fluctuation of friction forces during the penetration procedure. To simulate this feel, we add a random quantity generated equally from the range of $[-\delta\mu, \delta\mu]$ to the friction forces, where μ is the coefficient of kinetic friction and δ is a variable that can be adjusted according to the roughness of tissues. Thus, the friction forces can be calculated by

$$F_f(x,v) = -vv(2\pi rx) + random(-\delta\mu, \delta\mu), \tag{2}$$

where v and r are the velocity and radius of the needle, respectively; v is the viscosity of related tissue, and x is the displacement of needle penetration into the related tissue.

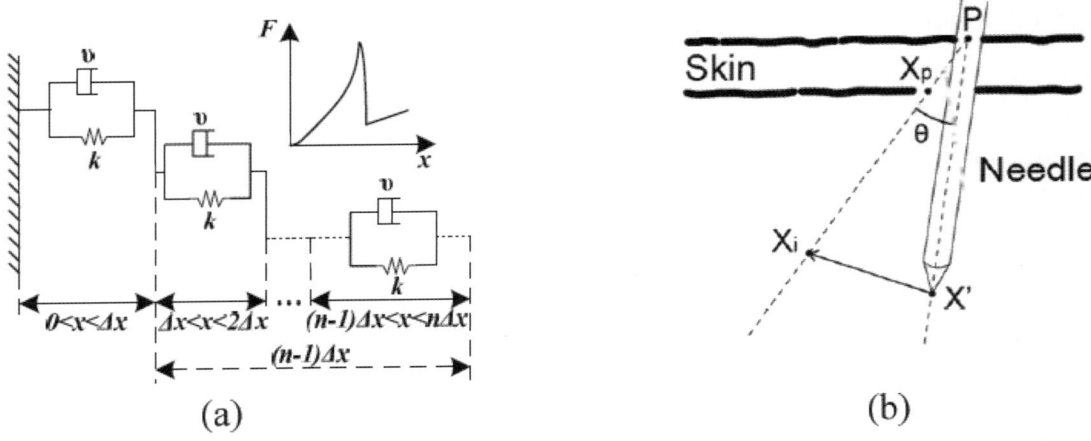

(a)   (b)

Figure 1. Haptic rendering: (a) incremental viscoelastic model and (b) path constraint force/ torque modeling.

3.2.3 Cutting Force Modeling

The cutting force is caused by collisions with and puncture of the interior structures of internal organs. Different organs have different models for cutting forces. For example, according to Okamura, Simone, and O'Leary (2004), cutting force can be approximated to a constant force F_c in the liver region. In the skin, adipose tissues, and *muscle*, we neglect cutting force because they do not contain such a substantial number of internal structures.

3.2.4 Needle Path Constraint

To maintain the insertion path, our system provides force feedback and torque feedback to prevent users from advancing or rotating the penetrated needle. The principles of force feedback and torque feedback are shown in Figure 1b, where P is the contact point of the skin surface and X_p is the puncture point of the skin. Therefore, the insertion path can be determined by the vector PX_p. At some moment, if the stylus tip deviates from the original path from X_i to X_o, the device will generate a resistant force F_{pc} and torque Q_{pc} to restrict the deflection based on the rotating axis and angle θ:

$$F_{pc} = m(d) * (PX_i - \frac{PX_p}{\|PX_p\|}) \|PX'\| \tag{3}$$

$$Q_{pc} = n(\theta) * \frac{PX' \times PX_i}{\|PX' \times PX_i\|}, \tag{4}$$

where d is the penetration depth and $m(d)$, $n(\theta)$ are some non-uniform amplification functions providing certain resistance even when d and θ are small. Note that haptic devices with 6-DOF force feedback (such as PHANToM Premium 1.5 High Force/6DOF) should be equipped to maintain the needle trajectory after puncturing the skin.

3.2.5 Force Composition

In addition to the feedback forces along the needle and the path constraint force and torque, we have to add a device weight compensation force (F_{wc}) and torque (Q_{wc}) to the total force F_{total} and total torque Q_{total}, respectively (Birtwisle & Bulpitt, 2007), since the structure and actuator mass of the haptic devices exert a downward force on the stylus tip even when no force is sent to the device. Consequently, the total force in every haptic frame is given by

$$F_{total} = \sum F_f + F_{pp} + F_c + F_{pc} + F_{wc} \tag{5}$$

$$Q_{total} = Q_{pc} + Q_{wc}, \tag{6}$$

where F_f is the total friction force caused by all penetrated tissues. F_{pp} is added during the puncture events; F_c is added when the needle penetrates into liver; and F_{pc}/Q_{pc} and F_{wc}/Q_{wc} are the path constraint force/torque added after skin penetration and the device weight compensation force/torque. All the force components can be calculated based on the aforementioned models.

4   Distributed Visual-Haptic Rendering

In traditional visual-haptic applications, haptic rendering and graphic rendering are usually carried out in different threads to fulfill the respective refresh requirement. Although this architecture has achieved some success in developing a VR-based surgical simulation system, we argue that it has several insurmountable difficulties in supporting a more complex environment where multiple haptic devices are involved.

First, the great disparity in update rate between graphic rendering and haptic rendering makes it easy to display an inconsistent state. That is, more computational resources are assigned to the haptic thread to fulfill its requirement of a high update rate (e.g., 1000 Hz). Meanwhile, the graphic thread cannot get enough resources to maintain a necessary update rate for continuous display, especially when (a) the graphic models are complicated and many objects are simultaneously moving or deforming in the virtual environment or (b) the computational resources are limited. Although we can utilize high-performance computers to deal with the second problem, cost efficiency, one of the most important advantages of VR-based surgical simulation, will be lost. In addition, as the field of biomechanical modeling matures, the models employed in surgical simulation should be more and more complex. More computational resources are needed to render them realistically and in real time. The problem becomes more complicated when multiple haptic devices are adopted in the applications.

In our framework, we propose a distributed visual-haptic framework (Figure 2). A client-server architecture is adopted to parallelize visual and haptic rendering in order to minimize performance overhead occurring in concurrently running haptic and visual rendering processes on the same machine. The visualization client is

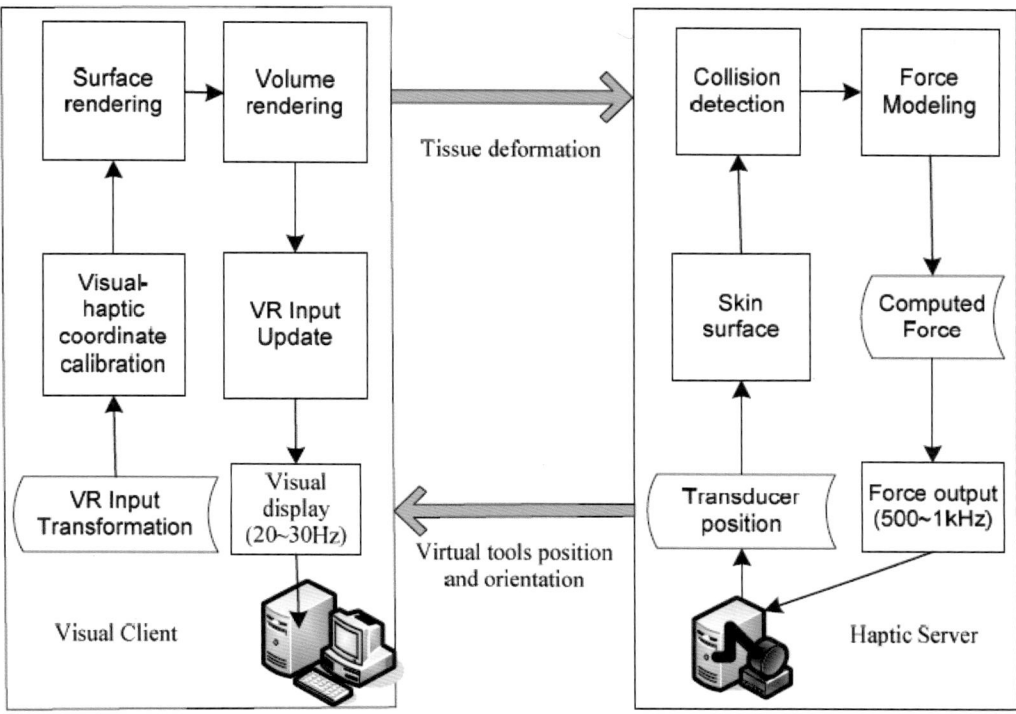

Figure 2. Client-server–based distributed visual-haptic architecture.

responsible for (a) collecting transformation states from the haptic devices to get the current position and orientation of the virtual tools and (b) rendering 3-D graphic models, including surface rendering and volume rendering. Simultaneously, the haptic servers can perform collision detection and haptic rendering at a relatively higher update rate. The other advantages of the proposed framework are that it enables the use of specialized computing architectures for both interfaces and achieves modularity during development.

5 Applications

5.1 Ultrasound-Guided Organ Biopsy Simulation

Ultrasound-guided biopsy is performed to find an abnormal area of tissue and guide its removal for examination. The success of this procedure is dependent on the correct alignment of the biopsy needle with the scanning plane of the ultrasound probe and with the target lesion, a skill that requires considerable training and practice to perfect. Both visual and haptic feedback are crucial to acquire the hand-eye coordination skills needed in the procedure. Although some work has been devoted to developing a simulation system for ultrasound-guided needle-insertion procedures (Vidal, Chalmers, Gould, Healey, & John, 2005; Zhu, Magee, Ratnalingam, & Kessel, 2007), most systems cannot provide interactive and realistic force sensation. Forest et al. (2007) presented a simulator with haptic devices for ultrasound examination and ultrasound-guided needle insertion. However, the needle is simulated by adopting a 3-DOF force-feedback device (SensAble PHANToM Omni), which cannot provide resistance torque force to maintain the original path of needle insertion.

We propose a VR-based simulation system for training ultrasound-guided needle insertion with realistic visual and haptic feedback based on the proposed haptic rendering framework. To provide the capability of two-handed training, we make use of two haptic devices to simulate a transducer and a needle. We use one SensAble PHANToM Omni and one PHANToM Premium 1.5 High Force/6DOF to simulate the transducer and the needle, respectively. The latter device can provide 6-DOF force feedback, which is necessary to maintain the needle's trajectory after puncturing the skin. We adopt different schemes to simulate the feel of the transducer and needle manipulations. For the transducer, we set a constraint on the skin surface to prevent the virtual transducer from passing through it. The user can feel a resistance force when the virtual transducer collides with the skin. Force modeling for needle insertion is implemented based on the integrated solution described in Section 3. Figure 3 shows the overall system.

Several parameters in our haptic model have to be calibrated before starting the training procedure, including the elastic and viscous coefficients of involved tissues, the coefficients of kinetic friction of the involved tissues, and the incremental displacement Δx for the incremental viscoelastic model. In the initialization step, we set these parameters according to the experimental results in the biomechanical references mentioned previously. Parameters are finely adjusted based on the practical feelings of experienced surgeons. Results are shown in Table 1. In addition, we provide an interface to assist users in adaptively adjusting these parameters according to patient-specific biomechanical properties.

We invited 12 novices and 2 experienced radiologists to evaluate our system; users found our realistic visual and haptic renditions improved the involvement in and training of needle insertion. Figure 4 shows the force-displacement relationship obtained in one complete needle-insertion training session. It is clearly observed that three force peaks occur at the puncture of skin, muscle, and liver capsule. The resistance forces after puncturing the liver are greater than those after puncturing skin because we have introduced an extra force to simulate the cutting of interior

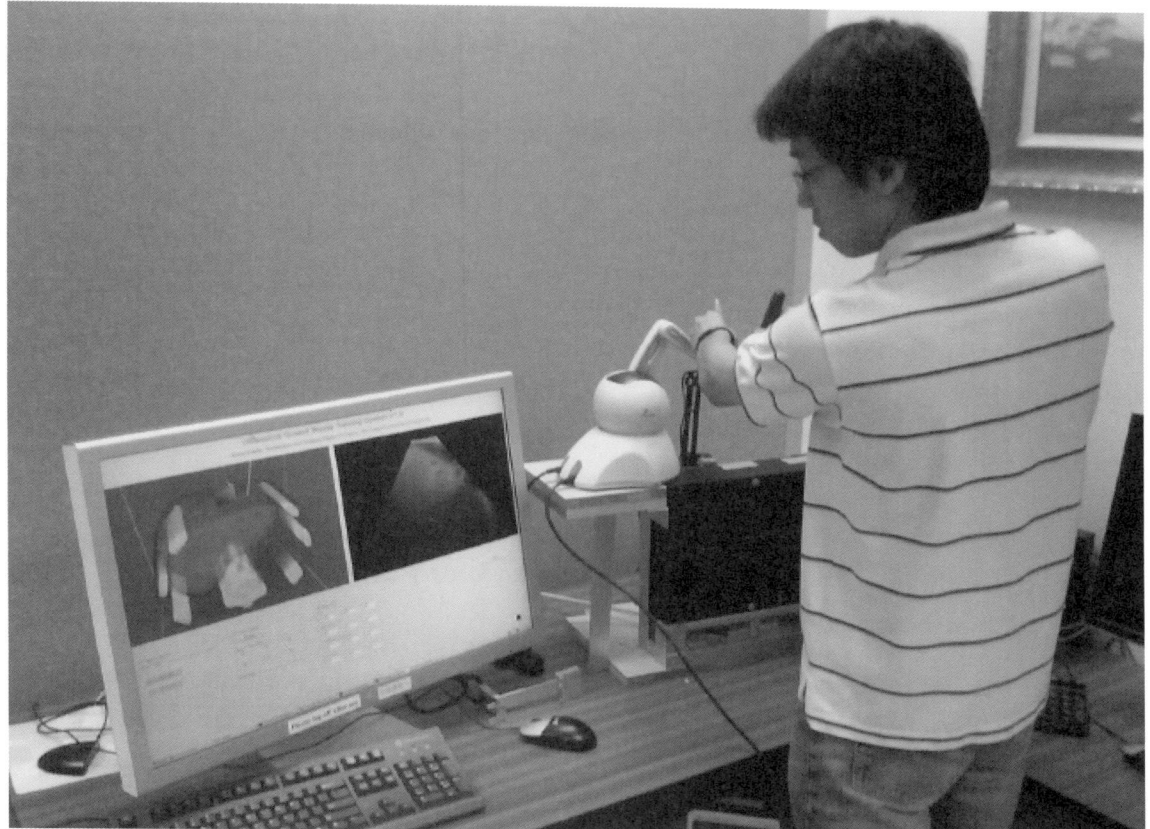

Figure 3. The ultrasound-guided needle-insertion training system.

Table 1. The Calibrated Parameters of Our Haptic Model

Parameter	k (skin)	v (skin)	μ (skin)	Δx (skin)	v (ad)	μ (ad)	k (muscle)
Values	0.46 N/mm	0.02 Ns/mm	0.18	0.1 mm	0.01 Ns/mm	0.1	0.19 N/mm
Parameter	v (muscle)	μ (muscle)	Δx (muscle)	k (liver)	v (liver)	μ (liver)	Δx (liver)
Values	0.03 Ns/mm	0.15	0.11 mm	0.42 N/mm	0.02 Ns/mm	0.58	0.12 mm

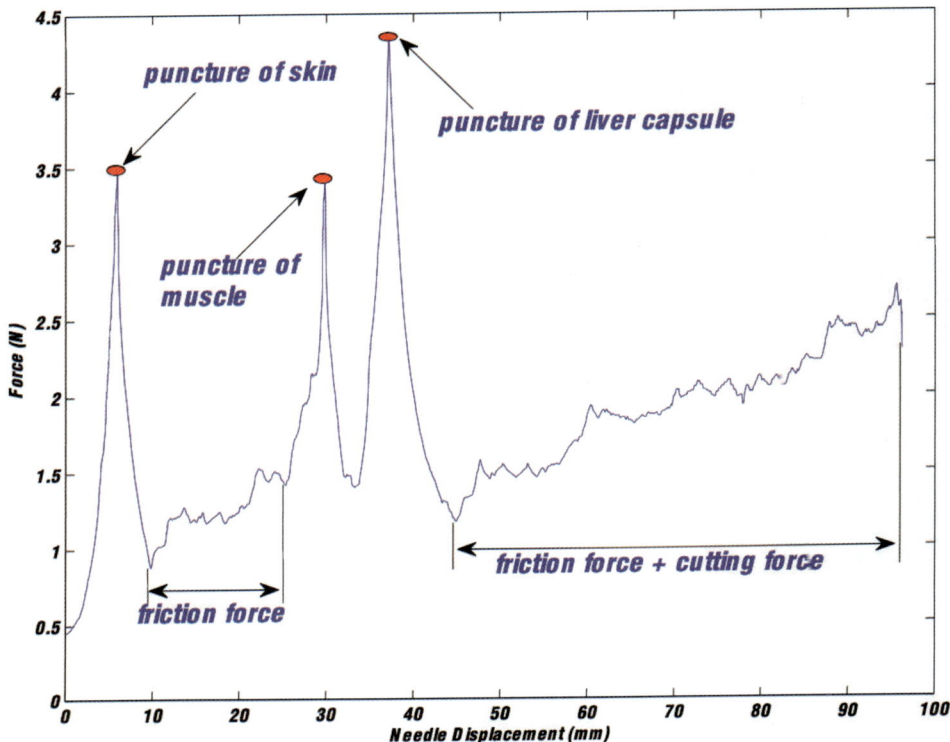

Figure 4. Force-displacement relationships obtained in one complete needle-insertion training session.

structures in the liver. It is demonstrated that our haptic modeling framework can provide users with a realistic force feeling throughout the needle-insertion procedure. Thanks to the distributed visual-haptic architecture, both haptic devices can maintain a high update rate of more than 1000 Hz and users can get a stable force sensation.

5.2 Chinese Acupuncture Training System

Another application of the proposed haptic modeling framework is a Chinese acupuncture training system. Chinese acupuncture is one of the key components of traditional Chinese medicine used to treat and prevent diseases by stimulating acupuncture points on the body. A target point can be stimulated by lifting, thrusting, and rolling a needle repeatedly with controlled depth and frequency. Guiding the needle to the destination precisely and manipulating it properly is of particular importance for achieving the desired therapeutic effects; the success of this process relies on correctly distinguishing the tissue based on the haptic experience with the needle.

Traditionally, acupuncture students can only practice on artificial mannequins or real patients. While an artificial mannequin provides very limited visual feedback and unrealistic force feedback, patient-based practice may impose potential complications. The incorrect identification of acupuncture points or incorrect manipulation may result in dizziness, pain, internal bleeding, and even long-term adverse effects.

We propose a VR-based system for Chinese acupuncture learning and training based on our framework. Acupuncture students can learn and practice acupuncture through a comprehensive virtual human model within a virtual environment. Our system presents the user with a force-feedback interface for needle-insertion training. Our system also provides informative visualization of acupuncture points of various related meridians where the collateral can be highlighted to guide the students during training. We use the stylus on a SensAble PHANToM device to simulate the Chinese acupuncture needle. The force acting on the needle is computed by integrating force components along the needle up to its tip based on our haptic models. The distributed visual-haptic architecture is adopted to connect a haptic workstation with a virtual workstation through the network. Having separated the haptic feedback generation system from the visualization system, the burden of force computations and haptic feedback delivery from the graphics workstation can be removed and thus results in increased system performance. Figure 5 shows the hardware configuration of the system.

In addition, because the acupuncture needle insertion is bidirectional, our resulting haptic model is capable of rendering the feedback force during a bidirectional needle movement at any position and angle within the virtual patient. Our system provides the functionality of recording the timing data from the needle-insertion practice. After a practitioner inserts a needle in the virtual body or pulls a needle out of the body, a training session is considered complete. The depth, velocity, and force can be recorded. The variation of force being simulated over time in our system has been demonstrated in Figure 6. An interactive calibration interface is set up so that haptic model parameters can be tuned by sliders. We have invited acupuncture practitioners from the Nanjing University of Traditional Chinese Medicine to tune the parameters according to typical needle-insertion experience and to subsequently justify the correctness of the haptic rendering presented through our simulation needling process. The generated force profiles are physically reasonable and agree with published data.

6 Conclusion

In this chapter, we propose an efficient and scalable framework for interactive haptic modeling and simulation of needle insertion and its application in training simulators

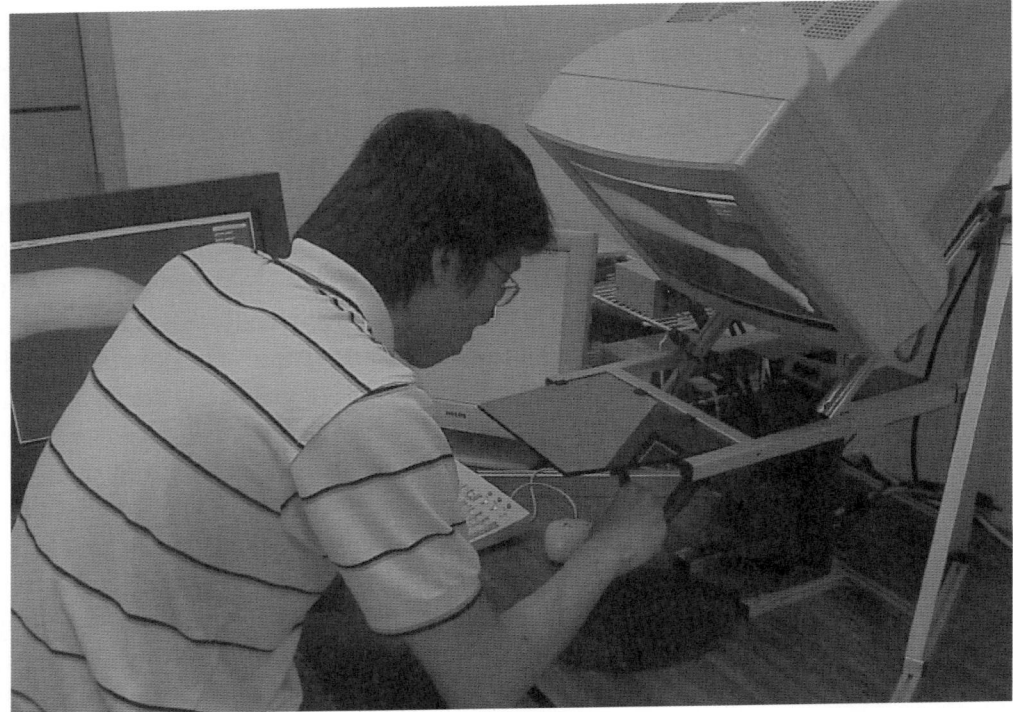

Figure 5. The Chinese acupuncture training and learning system.

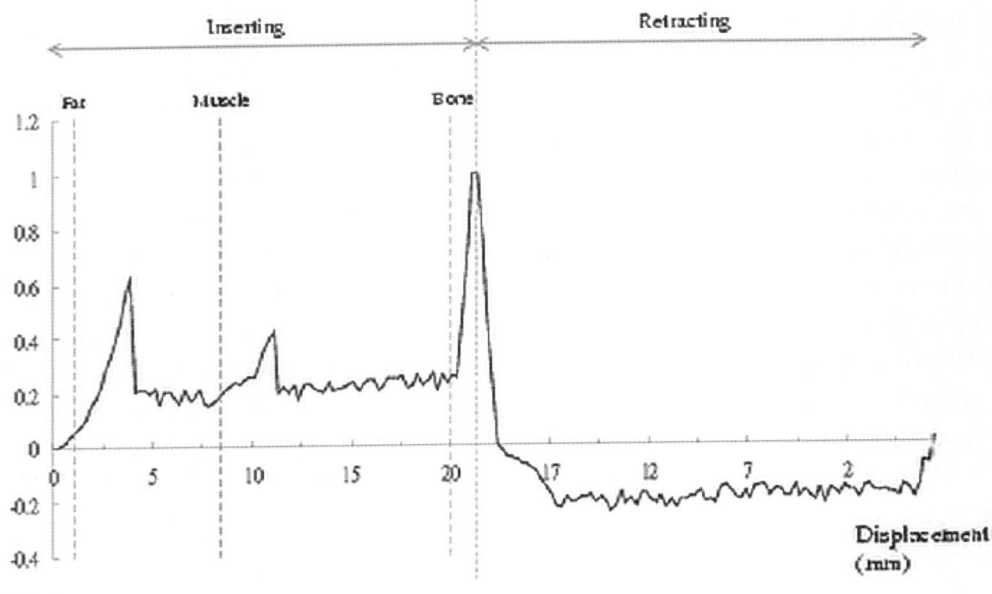

Figure 6. Force versus displacement for needle insertion of constant velocity (1 mm/s).

for percutaneous therapies. Our framework provides an integrated solution to model the various force components when the needle penetrates into skin, adipose tissues, muscle, and internal organs. In addition, after puncturing the skin, path constraint force and torque are simulated to maintain the initial insertion trajectory, which is important to ensure the task performance in some procedures. A client-server–based distributed visual-haptic framework is developed to parallelize visual and haptic rendering in order to support multiple haptic devices without degrading the whole system's performance. Our framework has been adopted to construct two medical training systems: an ultrasound-guided organ biopsy simulation system and a Chinese acupuncture training system. The success of these applications demonstrates the feasibility of our framework. To support the diverse requirement of various percutaneous therapy simulation systems, we currently focus on extending our framework to support surgical needles of different shapes and sizes (Abolhassani, Patel, & Moallem, 2007). In addition, we plan to introduce respiratory motion, needle bending, or related deformations into our framework in the future. More simulators should be implemented based on the proposed framework to further evaluate its effectiveness.

Acknowledgments

The work described in this chapter was fully supported by a grant from the Research Grants Council of the Hong Kong Special Administrative Region (Project No. CUHK4461/05M). This work is also affiliated with the Virtual Reality, Visualization and Imaging Research Center at the Chinese University of Hong Kong as well as the Microsoft-CUHK Joint Laboratory for Human-Centric Computing and Interface Technologies.

References

Abolhassani, N., Patel, R., & Moallem, M. (2007). Needle insertion into soft tissue: A survey. *Medical Engineering & Physics*, 29(4), 413–431.

Azar, F. S., Metaxas, D. N., & Schnall, M. D. (2000, June). A finite element model of the breast for predicting mechanical deformations during biopsy procedures. *Proceedings of the IEEE Workshop on Mathematical methods in Biomedical Image Analysis*, 38–45, South Carolina, USA.

Basdogan, C., De, S., Kim, J., Muniyandi, M., Kim, H., & Srinivasan, M. A. (2004). Haptic in minimally invasive surgical simulation and training. *IEEE Computer Graphics and Application*, 24(2), 56–64.

Birtwisle, M., & Bulpitt A. (2007). A 6DOF gravity compensation scheme for a PHANToM premium using a neural network. *Studies in Health Technology and Informatics*, 125, 43–48.

Brett, P. N., Parker, T. J., Harrison, A. J., Thomas, T. A., & Carr, A. (1997). Simulation of resistance forces acting on surgical needles. *Journal of Medical Engineering*, 211(4), 335–347.

Dang, T., Annaswamy, T. M., & Srinivasan, M. A. (2001). Development and evaluation of an epidural injection simulator with force feedback for medical training. *Studies in Health Technology and Informatics, 81,* 97–102.

Forest, C., Comas, O., Vaysière, C., Soler, L., & Marescaux, J. (2007). Ultrasound and needle insertion simulators built on real patient-based data. *Studies in Health Technology and Informatics, 125,* 136–139.

Gorman, P., Krummel, T., Webster, M. S., & Hutchens, D. (2000, January). A prototype haptic lumbar puncture simulator. *Medicine Meets Virtual Reality,* 106–109, Long Beach, CA, USA.

Kwon, D. S., Kyung, K. U., Kwon, S. M., Ra, J. B., Park, H. W., Kang, H. S., et al. (2001, May). Realistic force reflection in a spine biopsy simulator. *Proceedings of the 2001 IEEE International Conference on Robotics and Automation, IRCA 2001,* 1358–1363, Seoul, Korea.

Magee, D., Zhu, Y., Ratnalingam, R., Gardner, P., & Kessel, D. (2007). An augmented reality simulator for ultrasound guided needle placement training. *Journal of Medical and Biological Engineering and Computing, 45*(10), 957–967.

Niemeyer, G., Kuchenbecker, K. J., Bonneau, R., Mitra, P., Reid, A. M., Fiene, J., et al. (2004). THUMP: An immersive haptic console for surgical simulation and training. *Studies in Health Technology and Informatics, 98,* 272–274.

Okamura, A. M., Simone, C., & O'Leary, M. D. (2004). Force modeling for needle insertion into soft tissue. *IEEE Transactions on Biomedical Engineering, 51*(10), 1707–1716.

Vidal, F. P., Chalmers, N., Gould, D. A., Healey, A. E., & John, N. W. (2005, June). Developing a needle guidance virtual environment with patient specific data and force feedback. *Proceedings of Computer Assisted Radiology and Surgery,* 418–423, Berlin, Germany.

Zeng, J., Kaplan, C., Bauer, J., Xuan, J., Sesterhenn, I. A., Lynch, J. H., et al. (1998, February). Optimizing prostate needle biopsy through 3-D simulation. *Proceedings of SPIE Medical Imaging,* San Diego, CA, USA.

Zhu, Y., Magee, D., Ratnalingam, R., & Kessel, D. (2007, October/November). A training system for ultrasound-guided needle-insertion procedures. *Proceedings of Medical Image Computing and Computer-Assisted Intervention (MICCAI),* 566–574, Brisbane, Australia.

6 | # Virtual Exercise Environment for Participation and Adherence of People With Disabilities

P. Pat Banerjee, Ph.D. and Cristian J. Luciano
NIDRR Rehabilitation Engineering Research Center on Recreational Technologies and Exercise Physiology (Rectech), Department of Mechanical and Industrial Engineering, University of Illinois, Chicago

Abstract

This project develops and evaluates the use of virtual reality (VR) technology, together with appropriately adapted exercise equipment through an augmented reality interface, to create virtual exercise environments (VEEs). These environments allow persons with disabilities to exercise, train, and even compete with others. This chapter presents our current research on VEEs to facilitate participation and adherence using a rowing VEE as a prototype with open-source software and common off-the-shelf hardware components. The three specific aims of this project are to make exercise more enjoyable and less repetitious, to provide the means to easily set and track exercise participation and training goals, and to provide opportunities for individuals to engage in "virtual" competition with others. The responses of 33 participants comparing a three dimensional (3-D) immersive VEE with a regular monitor-based equivalent system are presented and analyzed.

Keywords: 3-D immersion, augmented reality, exercise environment, virtual reality

1 Introduction

Individuals who exercise regularly are healthier and tend to enjoy a better quality of life than those who are sedentary. This is especially true for persons with disabilities. For individuals not currently exercising, there are physical, attitudinal, and societal barriers to beginning and continuing a program of regular exercise. Recent empirical studies have demonstrated the importance of environment in influencing physical activity. A meta-analytic review of 19 studies by Humpel, Owen, and Leslie (2002) found that (a) accessibility of facilities, (b) opportunities for physical activity, (c) safety, (d) weather, and (e) aesthetics were leading determinants of participation in physical activity. While the Humpel et al. study reviewed research on the exercise habits of the general population, there is ample evidence that these factors are of even greater importance in determining participation in physical activity by people with disabilities. People with disabilities face a variety of personal and environmental barriers to participation in physical activity. Weather conditions such as excessive heat in summer or excessive cold and dangerous surfaces in the winter make outdoor exercise extremely hazardous for people with disabilities.

This chapter presents our current research on virtual exercise environments (VEEs) to facilitate participation and adherence. Exercising at home or in a fitness center presents numerous challenges for people with disabilities. Few fitness centers offer fully accessible opportunities for people with disabilities, and exercising at home can become boring for even the most dedicated exercise participant. Research has shown that individuals are more likely to engage in a program of regular exercise if the exercise is fun (i.e., their enjoyment outweighs their discomfort) and if they have a "partner" with whom they regularly exercise (Johnson, Rushton, & Shaw, 1996). For persons with disabilities, there are far fewer opportunities to exercise with a partner, and exercise may be more likely to be perceived as a chore than as an eagerly anticipated part of the day's activities (Heath & Fentem, 1997; Ravesloot, Seekins, & Young, 1998; Rimmer, 1999).

One approach to overcoming these barriers is to use technology to bring engaging, entertaining, and motivating exercise opportunities *to* people with disabilities. Specifically, this project develops and evaluates the use of virtual reality (VR) technology, together with appropriately adapted exercise equipment through an augmented reality interface, to create VEEs. These environments allow persons with disabilities to exercise, train, and even compete with others. The following are the three specific aims of this project:

1. Make exercise more enjoyable and less repetitious.
2. Provide the means to easily set and track exercise participation and training goals.
3. Provide opportunities for individuals to engage in "virtual" competition with others.

A prototype VEE with a wheelchair-accessible, augmented, VR-based cardiovascular rowing system has been developed not only to facilitate participation and adherence for people with disabilities but also with the intent of tapping into the potential for commercial or private manufacture, marketing, and distribution. FitCentric's NetAthlon, a commercial software application, has been adapted for the VEE along with WaterRower, a standard commercial rowing exercise machine. NetAthlon supports a variety of off-the-shelf and adapted exercise equipment. The prototype VEE uses Industrial Virtual Reality's ic3D augmented VR system.

The VEE was showcased at the 2006 National Institute on Disability and Rehabilitation Research (NIDRR) Rehabilitation Engineering Research Center Rectech State of the Science Conference on Exercise and Recreational Technologies for People with Disabilities in Denver. A number of people with disabilities used the VEE and were amazed at the potential of this technology. More follow-up studies are currently in progress.

2 Methods

Our approach makes extensive use of readily available technology used in gaming and entertainment products, making the development highly cost effective. Commercial trends continue to reduce the cost of the computing equipment required to create VEEs, making the acquisition of such equipment possible in the future even for people with modest means.

Each of the three specific aims of the project is described in more detail.

2.1 Make Exercise More Enjoyable and Less Repetitious

VR has been used in many different settings for persons with disabilities (Boian et al., 2002; Browning, Cruz-Neira, Sandin, & DeFanti, 1993; Jack et al., 2000; Johnson et al., 1996). Most of these uses have been associated with either rehabilitation or specific training. In addition, manufacturers of exercise equipment have in recent years added a sense of variety to their equipment using multimedia. Currently, the most common class of exercise equipment provides video monitors with different images meant to represent a variety of exercise locales, although synthetic-image VR experiences are beginning to be available.

VR may employ varying degrees of "immersion," ranging from a nonimmersive experience closely akin to watching television, to a synthetic interactive experience similar to a video game, and finally to an environment that attempts to create an experience as close to the "real" experience in three dimensions as possible (Browning et al., 1993). These "fully immersive" environments are rapidly moving from the computer graphics laboratory to the consumer. For this project, we provide VEEs for persons with disabilities for rowing waterways, lakes, or canals (Figure 1).

Using common off-the-shelf VR components to provide a more immersive and therefore a more realistic environment poses several significant questions for our testing. For example, what fraction of our population will be able to successfully use the system? What is the effect of the width of the field of view—does it only allow users to look straight ahead, or does the field extend to the peripheral vision of the user? This question is important, as there are instances in which individuals report adverse effects including nausea or dizziness (Lin, Duh, Abi-Rached, Parker, & Furness, 2002). These effects are sometimes referred to as "cybersickness" or "simulator sickness." The effect of simulator sickness is ameliorated with high-resolution graphics, and we are working to validate these claims for persons with disabilities.

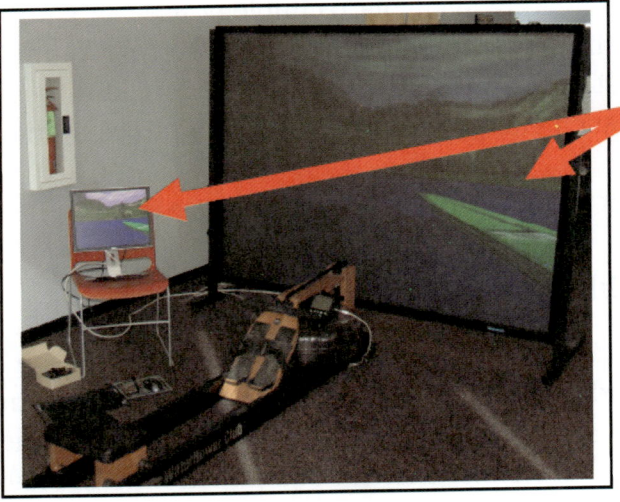

Comparing 3D Immersive Virtual Reality with LCD Display

Figure 1. Prototype #D Immersive and 2-D LCD-based Virtual Rowing Exercise Environment for Comparison Studies.

2.2 Easily Set and Track Exercise Participation and Training Goals

Other technology needed to create the VEE builds on engineering and technology efforts designed to provide feedback to support development of motor skills (Figure 2) and to provide an outlet for creative/aesthetic expression for persons with severe disabilities. The work generally involves a set of sensors that can detect even very small motions and the translation of these detected motions to control the auditory and visual environment. An early prototype project known as CARE HERE (Creating Aesthetically Resonant Environments for the Handicapped, the Elderly and Rehabilitation) has been developed by people in Europe and Australia.

This project adds sensors to synchronize the rowing environment. The visual field changes in proportion to the user's rowing machine belt pulling rate. We hypothesize that this capability will increase the user's engagement and enjoyment of exercise.

For exercise equipment that has been appropriately adapted, we are able to significantly improve the quality of the VEE by making it appear that the chosen route is being traveled at a rate consistent with the user's rowing speed. This level of interaction only requires that the exercise equipment be capable of reporting the rate at

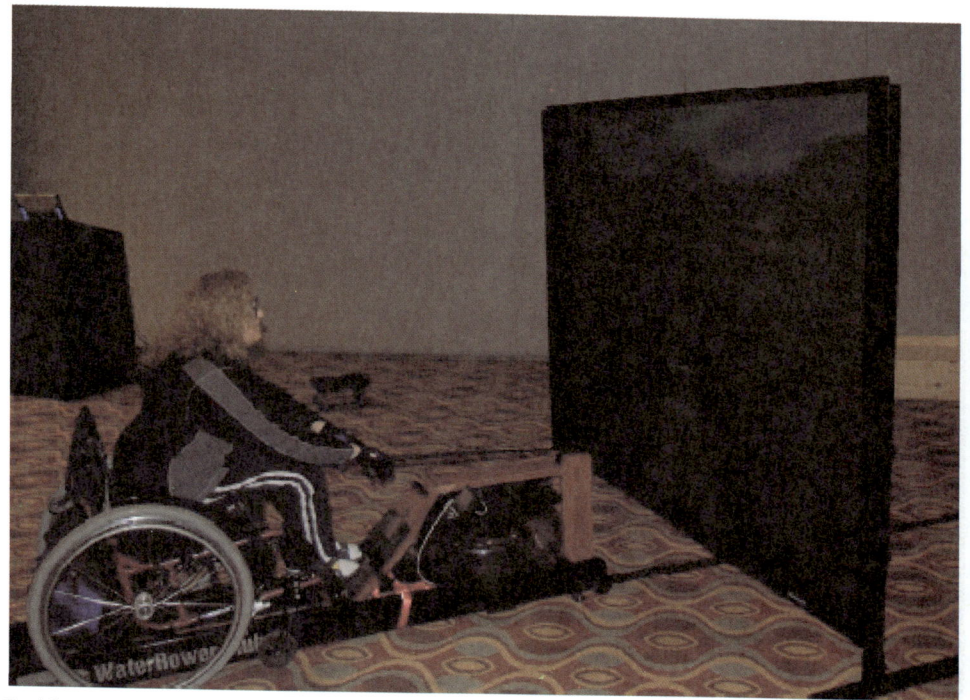

Figure 2. Motor skills enhancement used for persons with disabilities using a rowing VEE.

which the user is rowing. The second step in making the exercise environment more realistic is to provide force feedback to the exercise machine. For example, rowing can be made harder for an upstream course than a downstream course by changing the water resistance in the augmented reality simulation. Adding more water in the WaterRower container increases the water resistance and vice versa.

Those sharing VEEs may have a widely varying range of motor abilities and fitness. In order to allow individuals to easily share VEEs, each participant's motor ability and fitness has to be normalized to create a "level playing field." This normalization is based on the user's prior performance on the same or similar courses. When desired by the user, normalization can be disabled in order to provide a more direct measure of exercise progress.

2.3 Provide Opportunities for Individuals to Engage in "Virtual" Competition

We have begun to experiment with collaborative VEEs. Most of us are more likely to keep an exercise date if we make that date with a friend. However, individuals with disabilities may find it inconvenient, or difficult, to meet someone for this purpose. Transportation is one of the most frequently reported barriers to community participation for people with disabilities, making shared exercise programs problematic. Moreover, individuals may live a considerable distance from the person or persons with whom they wish to exercise.

We are developing a means for individuals to share VEEs. For example, two people who live in different cities could arrange to "ride" a particular rowing path.

Figure 3. Competition in virtual rowing.

Although each person is actually using an appropriately equipped device in his or her home, they can share the same VEE experience. In addition to the chosen path being represented to each person, an "avatar" of the other person also appears in the VEE, just as if they were rowing along the path together. An "avatar" is an on-screen image that represents the user in a VEE. Other users see the avatar as the representation of the user in the environment. The avatar represents a different colored rowing machine as selected by the user. For example, in our prototype example shown here, the user is represented as a red rowing machine while the competitor is shown by a green rowing machine. The relative position of the avatars—whether one rower is ahead of the other or the rowers are side by side—will reflect the relative rowing speed of the two riders, just as if they were rowing together on an actual lake or river.

3 Overall Architecture

This project represents a step toward developing an open-source Remote Exercise and Game Architecture Language (REGAL) system for Rectech. The main application of REGAL will be in the domain of remote recreation and rehabilitation, thereby alleviating the need for therapists to travel for routine monitoring of their patients. Additionally, the open source will facilitate collaboration among researchers and will lead to an alternative platform to compare and contrast the ingredients in some of the proprietary technologies being promoted by some private companies.

REGAL consists of (a) a baseline architecture and (b) a virtual augmented reality architecture on top of the baseline architecture. The design philosophy of the baseline architecture is based on Electro architecture.

Electro is an application development environment designed for use on cluster-driven tiled displays, VR systems, and desktop workstations. Electro is based on the MPI (message-passing interface) process model and is bound to the Lua programming language. With support for 3-D graphics, 2-D graphics, audio, networking, and input handling, Electro provides an easy-to-use scripting system for interactive applications spanning multiple hosts and a variety of displays. Using a set of scripts, Electro applications are iteratively created. It begins with a basic scene, adds a scene hierarchy, makes the scene interactive, and adds an intelligent camera. Electro is based on OpenGL (Open Graphics Language) and requires an NVIDIA GeForce FX (or better) or ATI Radeon 9600 (or better) graphics card.

The virtual augmented reality architecture on top of the baseline architecture is described by means of the prototype rowing VEE. The technical details of the prototype rowing VEE are provided next.

The Rowing VEE can be operated in three modes:

1. *Seat on wheels*, for participants who do not need wheelchairs
2. *Wheelchair on floor*, for participants whose wheelchair can climb on top of the rowing machine
3. *Wheelchair on ramp*, for participants whose wheelchairs cannot directly climb on the rowing machine

The software and hardware details are provided next.

3.1 Software

FitCentric Technologies Inc.'s (Montclair, California) NetAthlon (http://www .fitcentric.com/html/netathlon.htm), a commercial software application, has been adapted for the VEE. NetAthlon supports a variety of off-the-shelf and adapted exercise equipment. Rate of travel through the synthetic environment is proportional to the rate at which the rowing exercise equipment is operated. The current design of NetAthlon is geared more to support competition than to support shared exercise. Using a local area network or the Internet, NetAthlon permits rowers using fitness machines to compete against other rowers on fitness machines located elsewhere. Another FitCentric software product, UltraCoach, provides a customizable fitness data management system. NetAthlon uses rule-based training plans from workout data. The rules are intended to guide the user based on specific situations to keep the content interesting. It provides graphics in (a) a "first-person mode," in which the user's head bobs up and down, (b) a "third-person mode," in which a camera follows the user, and (c) a TV mode, in which the system changes camera shots like a TV show.

Proprietary solutions such as NetAthlon are not suitable for research and development because the architecture is not open. Lack of open standards hampers content development. Open standards for immersive VEEs are needed. Using lessons from CSAFE (Communications Specification for Fitness Equipment; http:// www.fitlinxx.com/CSAFE/), we are investigating open standards to provide an opportunity for multiple entities to build standardized VR content. The research on open standards is based on analyzing current technological components from (a) VR exercise software (e.g., FitCentric's NetAthlon); (b) 3-D courses (e.g., FitCentric courses); (c) exercise equipment such as the rowing machine from WaterRower (Warren, Rhode Island, http://www.waterrower.com/), which provides good initial compatibility with CSAFE and NetAthlon; (d) advanced graphics using Open Inventor Application Programmer Interface (API); (e) serial and/or wireless communication between the exercise equipment and the VEE; and (f) interface standards to projectors, big screens, and polarized glasses.

3.2 Augmented Reality Hardware

WaterRower, a standard commercial rowing exercise machine, has been adapted for the augmented reality user interface. Our prototype VEE uses Industrial Virtual Reality (Westmont, IL) ic3D system consisting of a 3.60-GHz, 2-GB RAM Intel Xeon Processor; NVIDIA ForceWare 3-D Stereo Driver; circular polarized glasses; two DLP projectors stacked on top of each other so that the images overlap (Figure 4); circular polarizers; and a 60-in. × 80-in. rear-projection screen from Stewart Filmscreen Corporation (Torrance, California). The rear projection screen is coated with a special polarization preserving material. These are custom made and coated with "Disney black diffusion film." The performance of such a material is superior because of its minimal stereo cross talk and because of its contrast qualities.

The prototype ic3D system is based on GeoWall technology (Vaidyasubramanian, Nayak, & Lopez, 2003). Two stacked InFocus LP530 DLP projectors have been used so that the images overlap (Figure 4). Circular polarization is used so that stereo is maintained even when viewers tilt their heads. Inexpensive passive stereo glasses have been used. A ForceWare NVIDIA 3-D Stereo driver facilitating full-screen stereo

Figure 4. Dual-stacked projectors in ic3D system based on GeoWall technology.

viewing of many Direct3D- or OpenGL-based applications has been used. NVIDIA ForceWare is a cheaper alternative to Christie's Active to Passive 3-D Converter (AP Converter). The update speed of the 3-D environment is synchronized and is proportional to the rate of the rowing paddle movement. Table 1 summarizes the hardware and software used.

4 Results and Discussion

The Rowing VEE was showcased at the 2006 NIDRR Rehabilitation Engineering Research Center Rectech State of the Science Conference on Exercise and Recreational Technologies for People with Disabilities in Denver. A number of people with disabilities used the VEE and were amazed at the potential of this technology. Figures 1 through 3 represent one such user at the State of the Science event.

A more formal study lasting for one session of about 30 minutes is currently being designed and is outlined here. The study participants exercise by pulling on a handle as in a standard rowing exercise machine. While performing the exercise, participants compete against computer-generated rowers. The primary purpose of this research is to assess the level of motivation and engagement by participants if a good VR environment is provided to them. The goal is to find ways to motivate people to participate in exercise. We aim to develop exercise technologies that are accessible to people with disabilities. As such, both able-bodied individuals and those with disabilities are participating in this research. The collected information will be used to assist in design changes on the resistance, aesthetics, and interface of the exercise device and virtual environment.

Consequently, two study groups are currently being designed: one consisting of able-bodied participants and the other consisting of people with disabilities. A questionnaire incorporating measures of satisfaction and other descriptive data for this initial cohort is used to gather information needed for the design and development

Table 1. Summary of Hardware and Software Used

Hardware	ic3D passive stereo system with a 3.60-GHz, 2-GB RAM, Intel Xeon Processor; NVIDIA ForceWare 3-D Stereo driver; circular polarized glasses; two DLP stacked projectors; WaterRower rowing machine
Software	FitCentric NetAthlon rowing software; Open Inventor API for VR modeling

process. Participants are asked to evaluate the perceived quality and value of the exercise environment. The major steps of the study are outlined next.

Step 1: First, an inclusion-exclusion criteria test is applied through a questionnaire. Eligible participants must

- not have had a traumatic injury to their shoulders, arms, or hands in the past;

- not have low or high blood pressure. The participant's blood pressure prior to participation is measured to ensure it is within normal levels (140–100/90–50 mmHg);

- not be receiving medical treatment for pressure sores or respiratory problems (asthma, etc.);

- not have any heart, lung, or other chronic medical conditions that prevent cardiovascular workout;

- be 18 to 50 years old.

Step 2: The participants then fill out an informed consent document. Following this, an assistant demonstrates the use of the rowing device/VR environment in the following manner. The assistant sits down on the rowing seat, positions his or her feet in the foot holder, puts on a pair of VR glasses, grasps the rowing handle with both hands, and begins the exercise session. The assistant pulls the rowing handle toward the chest while pushing the feet against the foot holder. The assistant demonstrates how the rowing device controls movement in the VR environment. The session lasts about 10 minutes. The assistant then takes off the VR glasses and gets up from the rowing seat.

Step 3: The participant then tries out the rowing device/VR environment. After about 10 minutes, the participant ends the session and fills out a questionnaire. If the participant has a physical disability, all steps will be the same as for other participants except for the following:

1. The rowing seat is removed and is replaced by a wheelchair to directly access the rowing device handle. Two options are pursued: If the wheelchair has enough clearance at the bottom to slide on top of the rowing machine guide bars, then it is directly used. If the wheelchair does not have enough clearance, then a ramp and platform is used.

2. The assistant demonstrates the use of the rowing device/VR environment using a wheelchair. It is important that the brakes on the wheelchair are in a locked position during the session so that the participant is stable.

Prior to, during, and following the exercise sessions, subjects are told to note the following conditions:

1. An unusual heartbeat, such as skipped beats or a very rapid pulse
2. Dizziness, light-headedness, feeling off balance, shortness of breath, nausea, or pain in any part of the body
4. Any loss of vision

The presence of any of these conditions is taken as an indication to terminate the subject's participation in the exercise.

Step 4: The last step consists of a postexperiment participant questionnaire based on certain relevant features from the design framework described in Witmer and Singer (1998). The first set of questions about the experiment, as listed in the appendix to this chapter, compare VR and desktop monitor experiences. The second set of questions, also listed in the appendix, attempts to capture background information on the participant, which helps us in normalizing the responses and in overcoming some biases.

Figures 5 through 9 show the results based on 33 participants in our Institutional Review Board–approved study comparing the described 3-D environment with a VR system using a regular monitor.

It is clear from the results that on most counts the 3-D immersive environment was better appreciated by the participants. In most cases, the standard deviation of the responses for the regular monitor was higher than that of the VR system responses,

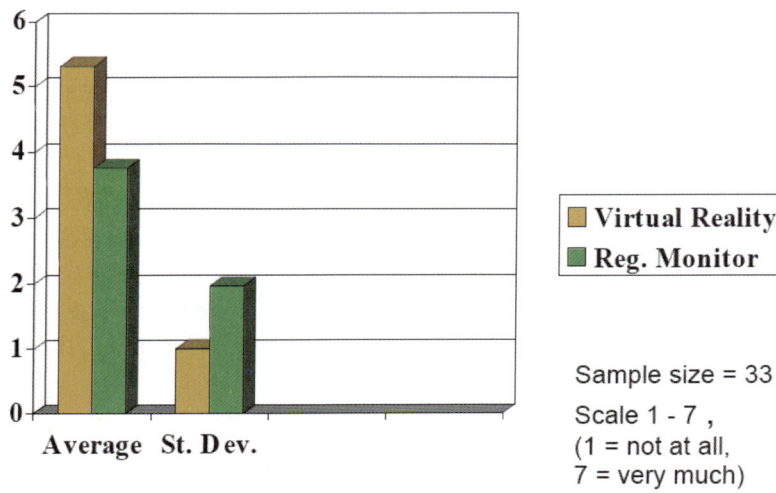

Figure 5. Responses for "How responsive was the environment to actions that you initiated?"

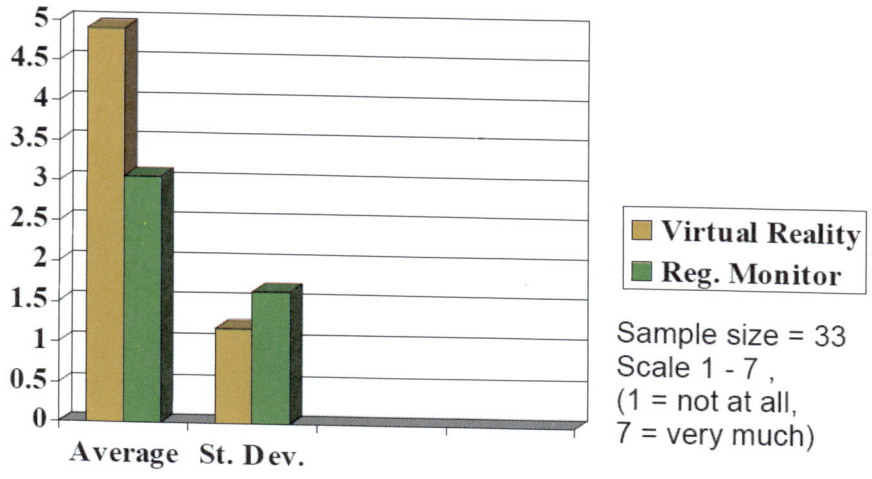

Figure 6. Responses for "How natural did your interactions with the environment seem?"

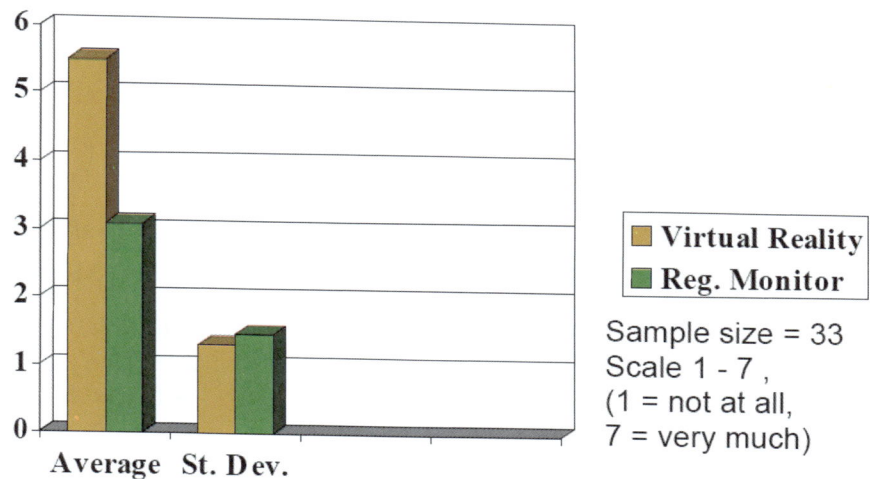

Figure 7. Responses for "How much did the visual aspects of the environment involve you?"

indicating that the responses for the VR system were not only better but also more consistently better.

5 Conclusion

The development of open standards for immersive VEE is an ongoing effort using the prototype rowing VEE. Immersive virtual environments are still in their infancy because compelling imagery and content creation for 3-D immersive environments is

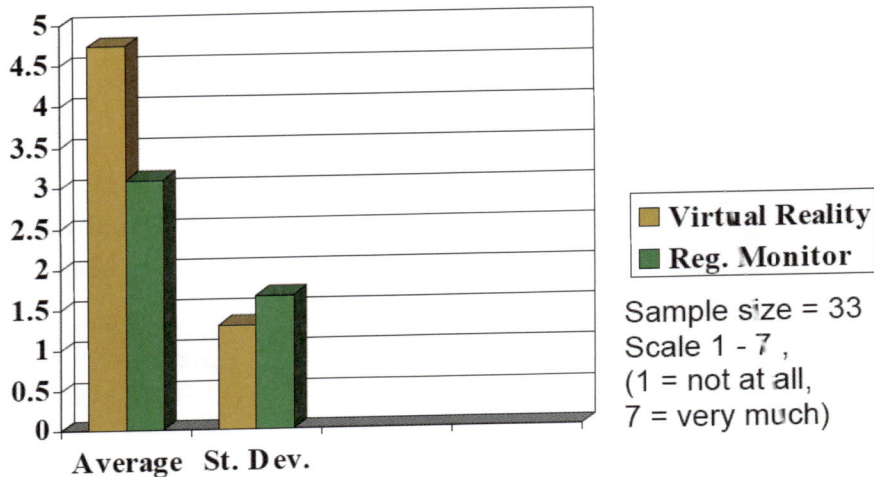

Figure 8. Responses for "How much did your experiences in the virtual environment seem consistent with your real-world experiences?"

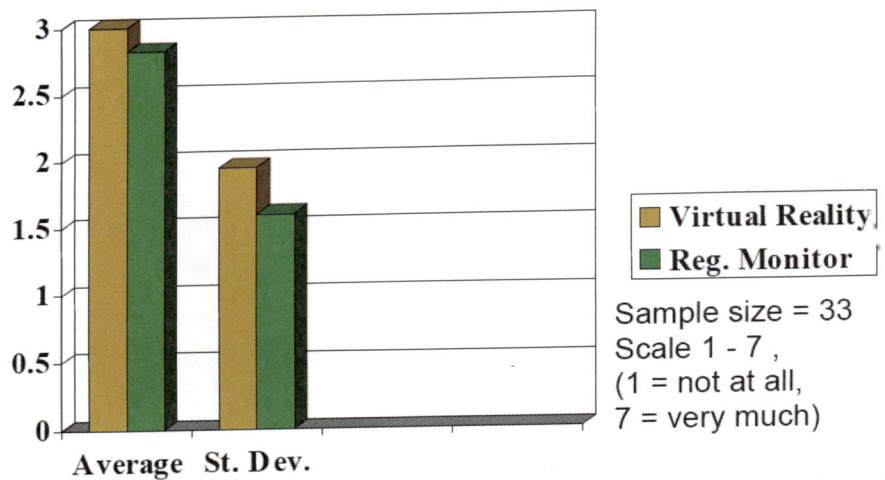

Figure 9. Responses for "How much did the visual display quality interfere or distract you from performing assigned tasks or required activities?"

a topic of active research and will take a few years to evolve. The nonimmersive content is more mature, but a direct conversion of nonimmersive to immersive content, as we have done here, is the very first step. Engaging immersive content will have to be developed independent of conversion from nonimmersive to immersive, and we hope to undertake this as a future research topic.

A further goal for wheelchair-based rehabilitation is to study the impact of VR and a companion rower on the cardiovascular workout. For this, we need to develop an augmented reality interface for a pair of rowers to exercise side by side. This concept can also be extended for rowers at geographically dispersed locations but connected through a teleimmersive VR environment.

Acknowledgments

This study was sponsored in part by NIDRR grant H133E020715 to Rehabilitation Engineering Research Center on Recreational Technologies (RERC Rectech) at the University of Illinois, Chicago. The authors wish to acknowledge some of the ideas given to us by Jim Rimmer and Bill Schiller that helped us write this chapter. Brian Faye helped with the design of the Institutional Review Board approval process documentation for our ongoing study.

References

Boian, R., Sharma, A., Han, C., Merians, A., Burdea, G., Adamovich, S., et al. (2002, January 23–26). *Virtual reality-based post-stroke hand rehabilitation.* Paper presented at the Tenth Annual Medicine Meets Virtual Reality Conference, Newport Beach, CA.

Browning, D. R., Cruz-Neira, C., Sandin, D. J., & DeFanti, T. A. (1993). *The CAVE projection-based virtual environments and disability.* Paper presented at the First Annual International Conference, Virtual Reality and People With Disabilities, San Francisco, CA.

Heath, G. W., & Fentem, P. H. (1997). Physical activity among persons with disabilities—a public health perspective. *Exercise and Sport Sciences Reviews, 25,* 195–234.

Humpel, N., Owen, N., & Leslie, E. (2002). Environmental factors associated with adults' participation in physical activity: A review. *American Journal of Preventive Medicine, 22*(3), 188–199.

Jack, D., Boian, R., Merians, A., Adamovich, S., Tremaine, M., Recce, M., et al. (2000, November 13–15). *A virtual reality-based exercise program for stroke rehabilitation.* Paper presented at the ASSETS 2000: Fourth ACM SIGCAPH Conference on Assistive Technologies, Arlington, VA.

Johnson, D. A., Rushton, S., & Shaw, J. (1996, July). *Virtual reality enriched environments, physical exercise and neuropsychological rehabilitation.* Paper presented at the First European Conference on Disability, Virtual Reality and Associated Technologies, Maidenhead, UK.

Lin, J. J.-W., Duh, H. B. L., Abi-Rached, H., Parker, D. E., & Furness, T. A. (2002, March). *Effects of field of view on presence, enjoyment, memory, and simulator sickness in a virtual environment.* Paper presented at the IEEE Virtual Reality Conference, Orlando, FL.

Ravesloot, C., Seekins, T., & Young, Q. R. (1998). Health promotion for people with chronic illness and physical disabilities: The connection between health psychology and disability prevention. *Health Psychology, 5,* 76–85.

Rimmer, J. H. (1999). Health promotion for people with disabilities: The emerging paradigm shift from disability prevention to prevention of secondary conditions. *Physical Therapy, 79*(5), 495–502.

Vaidyasubramanian, C., Nayak, A., & Lopez, B. (2003). How to put together a Geowall. Retrieved on November 4, 2009, from the Electronic Visualization Laboratory, University of Illinois at Chicago, http://www.evl.uic.edu/cavern/agave/docs/

Witmer, B. G., & Singer, M. J. (1998). Measuring presence in virtual environments: A presence questionnaire. *Presence: Teleoperators and Virtual Environments, 7*(3), 225–240.

Appendix

Questionnaire

The first set of questions:

1. How responsive was the environment to actions that you initiated (or performed)?
2. How natural did your interactions with the environment seem?
3. How much did the visual aspects of the environment involve you?
4. How much did your experiences in the virtual environment seem consistent with your real-world experiences?
5. How much did the visual display quality interfere or distract you from performing assigned tasks or required activities?

The second set of questions:

1. Do you easily become deeply involved in movies or TV dramas?
2. How frequently do you find yourself closely identifying with the characters in a story line?
3. Do you ever become so involved in a video game that it is as if you are inside the game rather than moving a joystick and watching the screen?
4. Do you ever become so involved in a daydream that you are not aware of things happening around you?
5. How well do you concentrate on enjoyable activities?
6. Have you ever gotten excited during a chase or fight scene on TV or in the movies?

7 Telerobotics Technologies for Virtual/Remote Manufacturing

D. J. Lee

Department of Mechanical, Aerospace and Biomedical Engineering,
University of Tennessee, Knoxville

Abstract

This chapter introduces recent ideas and results of telerobotics for use in virtual and remote manufacturing. In particular, after some basic notions and concepts of telerobotics are explained, the following topics are discussed: (a) how to achieve motion and power scaling between the real and virtual (or local and remote) environments; (b) how to implement virtual constraints to assist or guide human users or manufacturing tools; and (c) how to stably connect the real and virtual (or local and remote) environments when their communication medium is not perfect (e.g., Internet with delays).

Keywords: telerobotics, motion and power scaling, virtual constraints, Internet teleoperation, passivity

1   Introduction

The objective of this chapter is to introduce recent results and ideas in the field of telerobotics that would be useful for virtual (i.e., a real master device with a simulated slave robot) and remote (i.e., a real master device and a real slave robot, possibly geographically separated) manufacturing. The telerobotic field in general covers a vast research area. See Hokayem & Spong, 2006, for a recent survey on telerobotics. A typical telerobotic setup is shown in Figure 1, where a human operator manipulates a master robot (e.g., a haptic joystick) to command the slave robot's mechanical behavior (e.g., motion, forcing) while perceiving the external force experienced by the slave robot (e.g., contact force in an assembly task). The controller resides between the master and slave environments to achieve this. If the master and slave environments are connected via some imperfect communication (e.g., the Internet), this communication block may also be included in the controller block in Figure 1. This telerobotic system framework is promising for many important practical applications: telesurgery, remote nuclear waste management, cell manipulation, and space operations, to name a few. Here, if the slave environment is located in the real world, these applications may be thought of as similar to those in remote manufacturing. If the slave environment is a virtual one simulated in a computing station, these contexts become relevant for virtual manufacturing applications.

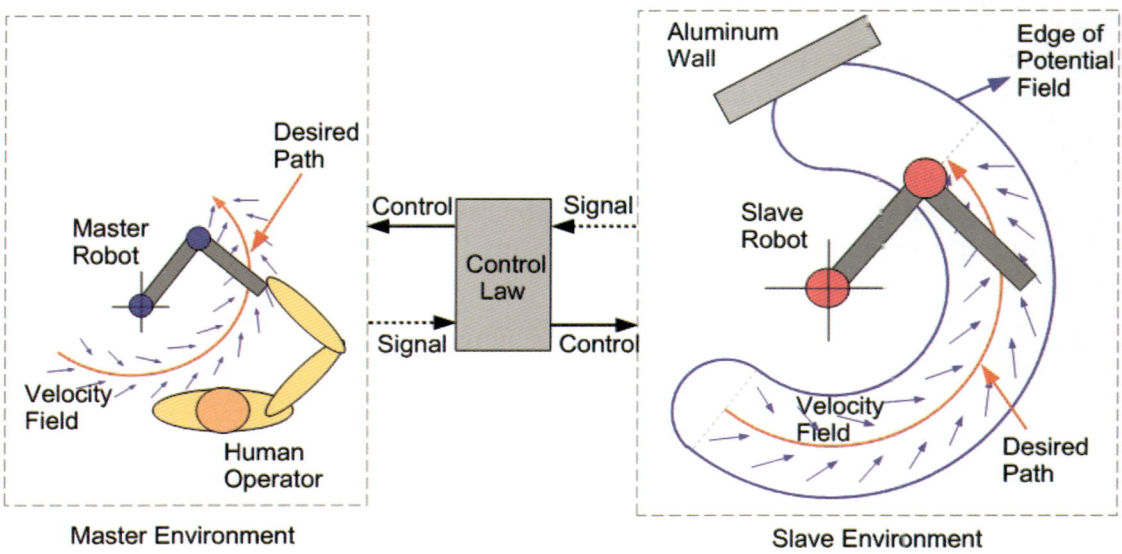

Figure 1. Telerobotic system consisting of master and slave robots (from Lee & Li, 2005).

The simulation of the virtual environment itself poses its own technical issues and challenges. For example, the usual telerobotic system requires force feedback that in turn demands a very fast update rate (e.g., 1 kHz) from the simulation. How to achieve this fast simulation while achieving perceptually convincing realistic physics of the virtual world is a fundamental problem in haptics (e.g., Barbič & James, 2008); however, we will not cover it in this chapter. Rather, in this chapter, we assume that, for virtual manufacturing applications, the simulation in the virtual environment has rapid response so that the slave system and its environment can be approximated reasonably well by a continuous system.

The master and slave robots can be modeled by the following mathematical expressions (i.e., multi-degree-of-freedom [DOF] Lagrangian dynamics):

$$M_1(q_1)\ddot{q}_1 + C_1(q_1,\dot{q}_1)\dot{q}_1 + g_1(q_1) = \tau_1 + f_1 \tag{1}$$

$$M_2(q_2)\ddot{q}_2 + C_2(q_2,\dot{q}_2)\dot{q}_2 + g_2(q_2) = \tau_2 + f_2 \tag{2}$$

where q_* is the configuration (or generalized coordinate, e.g., for a 6-DOF robot $[x, y, z]$-translation and pitch/yaw/role angles); M_* is the symmetric/positive-definite inertia matrix; C_* is the Coriolis matrix; g_* is the gravitational term; and τ_*, f_* are the control actions (e.g., given by actuator motors) and the external force (e.g., human force or contact force). Often, the gravitational term g_* can be locally canceled out. Also, if the operation is slow (i.e., $\|\dot{q}_*\|$ is small) and of limited workspace (i.e., the range of q_* is limited), these nonlinear robot dynamics can be approximated by the following linear robotic dynamics such that

$$M_1\ddot{q}_1 = \tau_1 + f_1 \, , \; M_2\ddot{q}_2 = \tau_2 + f_2 \, ,$$

for which more abundant results from telerobotics exist. With perfect communication, the feedback controls can be then designed for τ_1,τ_2 such that

$$\tau_1 = \tau_1(q_1,\dot{q}_1,q_2,\dot{q}_2,f_1,f_2) \, , \; \tau_2 = \tau_2(q_2,\dot{q}_2,q_1,\dot{q}_1,f_2,f_1) \, ,$$

where q_* is often measured by quadrature encoder and $\dot{q}_*$ by its numerical differentiation, while f_1,f_2 is measured by some force sensors. Acceleration feedback $\ddot{q}_*$ is generally not included unless additional accelerometers are deployed (which is usually not true in practice).

In many cases, we do not need to distinguish how the master side is controlled from how the slave side is controlled (e.g., slave behavior will also command that of the master device, while the human force is transmitted/perceived to the slave robot), and if so, we call the controlled telerobotic system bilateral (e.g., the functions τ_1, τ_2 above have the same structure [Lee & Li, 2005]). Usually, this bilateral control architecture provides position coordination (i.e., $q_1(t) \rightarrow q_2(t)$) and force feedback (or haptic feedback/force reflection: $f_1(t) \rightarrow f_2(t)$). We say the telerobotic system is *ideally transparent* if the controls τ_1, τ_2 achieve perfect position coordination and force reflection at the same time, regardless of human/slave forcing,

$$
\begin{pmatrix} F_1(s) \\ -V_2(s) \end{pmatrix} = \begin{bmatrix} H_{11}(s) & H_{12}(s) \\ H_{21}(s) & H_{22}(s) \end{bmatrix} \begin{pmatrix} V_1(s) \\ F_2(s) \end{pmatrix} = \begin{bmatrix} O & -I \\ I & O \end{bmatrix} \begin{pmatrix} V_1(s) \\ F_2(s) \end{pmatrix}
\tag{3}
$$

where $V_*(s) = L[\dot{q}_*]$ and $F_*(s) = L[f_*]$ with s and $L[\cdot]$ being the Laplace variable and the Laplace transform (Hannaford, 1989; Yokokohji & Yoshikawa, 1994). This ideal transparency is the best performance we can achieve from the telerobotic system, implying that a human operator can perceive exactly the same motion and force the slave experiences. How far the system departs from this ideal transparency can then be thought of as a performance measure.

This ideal transparency, in general, requires the four-channel control architecture (Lawrence, 1993), which requires both the kinematic feedback (i.e., $q_*, \dot{q}_*$) and the force sensing transmission (i.e., f_1, f_2) in the controls τ_1, τ_2. Although this four-channel architecture provides a performance as good as it gets, its stability is often not so robust, similar to the case of a very high-performance control system that requires a precisely tuned feed-forward control loop that can become easily unstable in the presence of model/parameter uncertainly and/or noise. Also, for this four-channel architecture, control-gain tuning may take some time.

In many practical cases in which the required level of performance is not so high (we presume this is true for many virtual or remote manufacturing applications), often the following simple proportional-derivative (PD, or spring-damper) type of control can provide an adequate level of performance (Lee & Huang, 2008b; Lee & Spong, 2006; Oboe & Fiorini, 1998):

$$
\tau_1 = -B(\dot{q}_1 - \dot{q}_2) - K(q_1 - q_2), \quad \tau_2 = -B(\dot{q}_2 - \dot{q}_1) - K(q_2 - q_1),
\tag{4}
$$

where B, K are positive-definite and symmetric damping and spring gain matrices. Here, the damping feedback term also can be replaced by just the local damping dissipation $-B\dot{q}_1$ and $-B\dot{q}_2$. Then, in steady-state, with some technical assumptions (e.g., Lee & Spong, 2006), we can show that (a) if $f_1 = f_2 = 0$, $q_1 \rightarrow q_2$ (i.e., position coordination) and (b) if $(\ddot{q}_*, \dot{q}_*) = 0$, $f_1 \rightarrow K(q_1 - q_2) \rightarrow -f_2$ (i.e., force reflection). Due to its simplicity, this PD control has been frequently used in practice. Also, as recently shown in Lee and Huang (2008b) and Lee and Spong (2006), this PD control, with a suitable gain setting or some extra passivity-enforcing subroutines, can still be used with communication impefectness (e.g., constant/varying delay, packet loss, time swapping). In this chapter, we will primarily focus on this PD control as a means to connect the master and slave robots (Section 4 provides more details).

Performance will not be important if we cannot enforce the stability of the closed-loop telerobotic system. For this stability problem, let us consider a more abstract representation of Figure 1: See Figure 2, where we have master-slave communication delays. Then the closed-loop teleoperator is a two-port system that interacts with the human operator (left side) and the slave environment (right side). The key aspect of telerobotics, which sets it apart from other control systems in terms of stability, is that feedback loops are created between this two-port closed-loop telerobotic system and the human operator (again, left side of Figure 2) and the slave environment (right side of Figure 2). Moreover, these human operators and slave environments are often unknown, unmodeled, and complicated that the (otherwise powerful) traditional model-based stability analysis tools (e.g., Nyquist criteria, Routh-Hurwitz criteria, etc.) are not applicable anymore.

The concept of passivity has been widely utilized to address this interaction-stability problem of telerobotic systems (e.g., Anderson & Spong, 1989; Hannaford & Ryu, 2002; Lee & Huang, 2008; Lee & Li, 2003; Lee & Spong, 2006; Niemeyer & Slotine, 2004; Stramigioli, van der Schaft, Maschke, & Melchiorri, 2002). That is, by

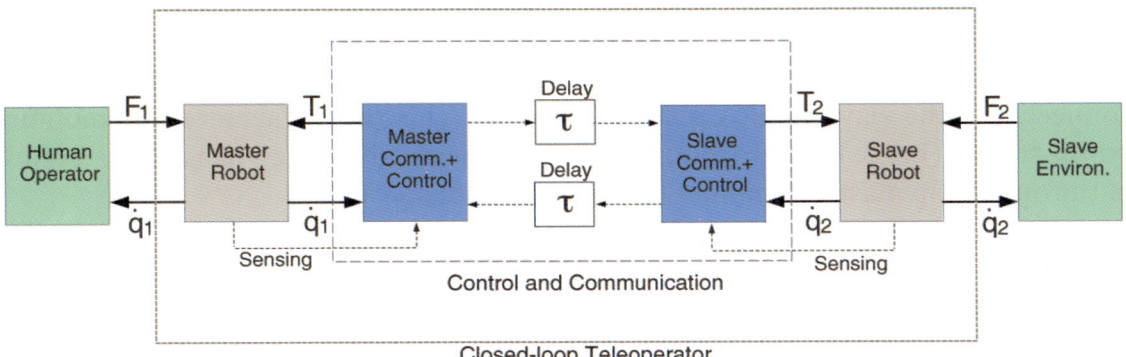

Figure 2. Telerobotic system consisting of master and slave robots (from Lee & Spong, 2006).

enforcing the (closed-loop energetic) passivity of the two-port such that for all $T \geq 0$, there exists a bounded $d \in \Re$ such that

$$\int_0^T [f_1^T \dot{q}_1 + f_2^T \dot{q}_2] d\tau \geq -d^2 \tag{5}$$

(i.e., the maximum extractable energy from the two-port telerobotic system is always bounded by d²) from the passivity theorem (Vidyasagar, 1993). The interaction stability with any passive humans and slave environments can be guaranteed, regardless how unknown, uncertain, or complicated (e.g., nonlinear/time varying) their dynamics are. Indeed, in many cases, both the human operator and the slave environment behave similarly to passive systems (e.g., passive human assumption [Hogan, 1989] or a docking task in which the slave environment can be modeled as a spring damper type wall, which is passive since it can only store or dissipate energy and cannot generate energy by itself). The above PD control (when the master-slave communication is perfect) ensures this two-port passivity of the closed-loop telerobotic system (since the spring and damping control elements and the open-loop master and slave robots are all intrinsically passive).

In the next section, we will discuss master-slave motion and power scaling, which are useful when the master and slave environments are of different scales.

2   Motion/Power Scaling

Suppose that the slave environment is of a drastically different spatial scale than that of the human operator (e.g., a slave environment for molecular manipulation, microassembly, or large civil-structure construction). In this case, directly connecting the (human-scale) master device and the (nano-, micro-, or large-scale) slave robot via some control (e.g., PD control) would not be useful due to the scale difference. It would be much more natural to first scale down (or up) the master side and then connect it to the slave robot via the PD control. This can be achieved by motion/power scaling (Lee & Li, 2003, 2005) while keeping the dynamics/passivity structure of the open-loop master dynamics (Equation 1) intact.

First, let us define a new scaled configuration

$$q_1' := S q_1,$$

where $S \in \Re^{n \times n}$ is a full-rank motion-scaling matrix. For a nano- or micro-scale application, we may consider using small S, whereas for a large-scale task, we might want to use a larger S. Then, by rewriting Equation 1 with respect to the new scaled configuration q'_1, we can achieve (with the gravitational term locally canceled out)

$$M'_1(q'_1)\ddot{q}'_1 + C'_1(q'_1,\dot{q}'_1)\dot{q}'_1 + g'_1(q'_1) = \tau'_1 + f'_1,$$

where $M'_1 = S^{-T} M_1 S^{-1}$, $C' = S^{-T} C_1 S^{-1}$, and $\tau'_1 = S^{-T}\tau_1$ (similar hold for f'_1). This motion-scaled master dynamic has the scaled configuration q'_1; thus, we can use the PD control (Equation 4) to directly connect it to the slave system (Equation 2) with matched spatial scale. Here, note that this new scaled dynamic with respect to q'_1 is merely a different representation of Equation 1 but with the same dynamic. For example, once the PD control is designed for τ'_1 as in Equation 4, the real control applied to the master robot (Equation 1) will be

$$\tau_1 = S^T \tau'_1 = S^T [-B(\dot{q}'_1 - \dot{q}_2) - K(q'_1 - q_2)],$$

while for the slave side we can use

$$\tau_2 = -B(\dot{q}_2 - \dot{q}'_1) - K(q_2 - q'_1).$$

Let us consider the control power of the spatially scaled dynamics of q'_1:

$$(\tau'_1)^T \dot{q}'_1 = (S^{-T}\tau_1)^T S\dot{q}_1 = \tau_1^T \dot{q}_1,$$

where S is invertible. This also holds for the human power (i.e., $(f'_1)^T \dot{q}'_1 = f_1^T \dot{q}_1$). This implies that the control power is not scaled down or up (although the motion has been). For example, in certain nano- or micro-scale applications, it would not be safe to directly connect the human control port to the slave control port. In some cases, the information transmitted from the slave environment to the human operator may contain insufficient power for the human to perceive it.

To match such a difference in power scales between the master and the slave environment (following Lee and Li [2005]), we power scale the entire motion-scaled master system with respect to q'_1 such that, with the slave dynamics (Equation 2) intact,

$$\rho[M'_1(q'_1)\ddot{q}'_1 + C'_1(q'_1,\dot{q}'_1)\dot{q}'_1 + g'_1(q'_1) = \tau'_1 + f'_1],$$

where $\rho > 0$ is a power-scaling such that if $\rho < 1$, the human power will appear to be attenuated to the slave environment (e.g., for nano- or micro-scale applications); and if $\rho > 0$, the human power will appear to be amplified to the slave environment (e.g., for a large-scale civil structure construction task). Here, this power scaling $\rho > 0$ is purely a mathematical operation that does not change any underlying physics of the master system. For example, when deriving the PD control (Equation 4), we can represent

$$\rho \tau_1' = -B(\dot{q}_1' - \dot{q}_2) - K(q_1' - q_2), \text{ that is, } \tau_1' = \frac{1}{\rho}[-B(\dot{q}_1' - \dot{q}_2) - K(q_1' - q_2)],$$

which can be recovered to τ_1. With this power scaling, ρ, instead of Equation 5, we can show the power-scaled two-port passivity of the closed-loop telerobotic system: for all $T \geq 0$,

$$\int_0^T [\rho f_1^T \dot{q}_1 + f_2^T \dot{q}_2] dt = \rho \kappa_1(T) - \rho \kappa_1(0) + \kappa_2(T) - \kappa_2(0) + \varphi_E(T) - \varphi_E(0) - \int_0^T \|\dot{q}_1' - \dot{q}_2\|_B^2 dt,$$

$$\geq -\rho \kappa_1(0) - \kappa_2(0) - \varphi_E(0) =: -d^2$$

where $\kappa_* = \frac{1}{2} \dot{q}_*^T M_* \dot{q}_*$ is the kinetic energy and $\varphi_E = \frac{1}{2}(q_1' - q_2)^T K(q_1' - q_2)$ is the coordination-error potential energy. When $q_1' = q_2$ with $f_2 = 0$ (or $f_1 = 0$, respectively), the "locked" inertia of the combined master and slave robots perceived by the human operator (or the slave environment, respectively) will be $M_1 + M_2/\rho$ (or $\rho M_1 + M_2$, respectively), whereas in the steady-state with no acceleration or velocity, we will have the scaled force reflection such that $\rho f_1 \rightarrow K(q_1' - q_2) \rightarrow -f_2$.

This motion/power scaling will be very useful for virtual or remote manufacturing applications with such scale differences between the master and slave environments, as we can change the motion scaling S and the power scaling ρ separately. For more details, refer to Lee and Li (2003, 2005), Goldfarb (1999), and Kosuge, Itoh, and Fukuda (2000).

3 Virtual Constraints

For some virtual or remote manufacturing applications, it may be useful to have virtual constraints or guidance to help humans perform various tasks. This may be accomplished by imposing some virtual dynamics locally for the individual master and slave environments. For example, we may render a virtual wall (i.e., virtual spring + virtual damping) in the slave environment (i.e., command the slave robot's actuator to generate a virtual contact force when the slave robot enters a region of a virtual wall) to prevent the slave robot from going outside of a certain region (e.g., the kidney-shaped region in Figure 1). Such a virtual wall (or a potential field with a damping injection [Khatib, 1986; Koditschek, 1991]) may also be used for collision avoidance (e.g., to protect a fragile object in the slave environment) or for virtual guidance to precisely align or locate a part for microassembly. Such issues can be studied under tool dynamics (Kosuge et al., 2000; Lee & Li, 2005) or virtual fixtures (Bettini, Marayong, Lang, Okamura, & Hager, 2004; Rosenberg, 1993). We can refer to them as virtual constraints.

To illustrate how this can be performed, let us consider the slave robot (Equation 2). Exactly the same procedure can be followed for the master side before the motion or power scaling of Section 2, since these virtual constraints will be rendered locally (i.e., for the master and slave sides individually). For the slave side, virtual wall–type virtual constraints can then be rendered by having the following control terms embedded in τ_2 such that

$$-B_w \dot{q}_2 - d\varphi_w(q_2) = -B_w \dot{q}_2 - \left[\frac{\partial \varphi_w}{\partial q_2^1}, \quad \frac{\partial \varphi_w}{\partial q_2^2}, \quad \cdots, \quad \frac{\partial \varphi_w}{\partial q_2^n} \right], \quad \text{if } q_2 \in W,$$

where B_w is the virtual damping, $\varphi_w(q_2) \geq 0$ is the potential function to define a certain desired potential field on the slave environment, q_2^i is the ith component of q_2, and $W \subset \Re^n$ is the region where this potential field is activated (e.g., outside of the kidney-shaped region in Figure 1). With the PD control (Equation 4), the total control (τ_2) will then become

$$\tau_2 = -B(\dot{q}_2 - \dot{q}_1) - K(q_2 - q_1) - B_w \dot{q}_2 - d\varphi_w(q_2),$$

which is a combination of the two control actions. Instead of using a single potential function φ_w, a concatenation of multiple spring-like potential fields can be adopted

$$-B_w \dot{q}_2 - K_{wi}(q_2 - q_{wi}), \quad \text{if } q_2 \in W_i,$$

where $W_i \subset \Re^n$ is a region in which we want the robot to experience repulsive or attractive spring action with the spring stiffness K_{wi} and the center point q_{wi} with i being the switching index. This spring concatenation may induce the problem of violation of passivity due to the spring potential energy jumping at the boundaries between W_i and W_j (Lee & Huang, 2008a; Liberzon & Morse, 1999). If the concatenation of the spring potential is smooth enough across W_i, the robot's motion is not so rapid, and the robot has a high enough level of intrinsic device damping, this issue of violating passivity would not be significant, and many passivity-enforcing control techniques (e.g., PD control [in Equation 4]) can be used enforcing the two-port passivity of the closed-loop telerobotic system.

Other forms of the virtual constraints are also possible. For instance, passive velocity field control (PVFC; Lee & Li, 2005; Li & Horowitz, 1999) can be used, as shown by the arrows in Figure 1, to guide the robot to follow a desired velocity field at each configuration point (i.e., at each point in Figure 1, the robot will follow the directions indicated by the arrows). This may be useful for some repetitive tasks to which the human operator does not necessarily want to pay close attention (e.g., moving the slave robot between the tool-exchange point and the real operation site). This PVFC also enforces passivity; thus, it can be used with other passivity-enforcing control techniques. See also Duindam and Stramigioli (2004) for passively following curves and Peshkin and Colgate (1999) for using nonholonomic virtual constraints (i.e., no-slip constraints of the wheels).

4 Passive Coupling Control

With the motion and power scaling and virtual constraints in place individually for the master and the slave, the next task that needs to be addressed will be how to connect the master and the slave robots with each other so that we can have some force reflection (i.e., $f_1 \to f_2$) and position coordination (i.e., $q_1 \to q_2$). For this, the most widely used technique is the PD control (Equation 4). The system performance will be mainly determined by how large the spring gain K can be. For example, the larger K is, the sharper the force reflection will be (i.e., with the same $q_1 - q_2$, there will be more acute force). With the larger K, the position error will be smaller with the same external force f_* and system dynamics (i.e., the same $\ddot{q}_*, \dot{q}_*$).

As in any control system, we cannot increase the control gain K indefinitely. There exists an upper bound for this K, until the closed-loop telerobotic system starts becoming unstable. This may be attributed to certain unmodeled dynamics. An alternative (and more interesting) explanation for this can be drawn from Brown and Colgate (1998) and Colgate and Schenkel (1997); that is, although the servo-loop rates for the master and slave are very fast (e.g., close to 1 kHz), if the control servo loop is digital (i.e., uses zero-order hold), the maximum K would be bounded by the celebrated virtual-wall passivity condition (Colgate & Schenkel, 1997): for the 1-DOF linear robot case,

$$B_{dev} > \frac{KT}{2} + B,$$

where B_{dev} is the unmodeled inherent device viscous damping (not incorporated in Equations 1 and 2), T is the digital control sampling rate, and B, K are control damping and spring gains. This condition states that if we want the two-port passivity of the closed-loop telerobotic system and a large spring control gain K, we need a fast sampling rate T and a large device-damping B_{dev}. Note that, in this case, increasing the control damping B will not help increase the bound for K.

In certain applications of virtual or remote manufacturing, the communication between the master and the slave may suffer from communication imperfectness such as varying delay, data loss, etc. This is particularly true if we use the ubiquitous Internet or wireless communication as the communication medium between the master and slave environments (e.g., virtual environment simulated in a remote server connected over the Internet) or if the distance between the master and slave sites is so great that the data transmission time itself is not negligible (e.g., a master site in a company headquarters and slave site in another country). In this case, the master-slave communication delay is not negligible and can easily make the closed-loop system unstable.

The most well-known approach for the time-delay problem in telerobotics is the scattering approach (Anderson & Spong, 1989), which has been refined in Niemeyer and Slotine (1991, 2004) with the introduction of the notion of wave transform. The key idea of this scattering technique is communicating its scattering (or wave) variable pair (s_*^+, s_*^-), which are defined for the master side such that

$$s_1^+ = \frac{N^{-1}}{\sqrt{2}}(\tau_1 + Z\dot{q}_1^d), \ \ s_1^- = \frac{N^{-1}}{\sqrt{2}}(\tau_1 - Z\dot{q}_1^d),$$

where $Z = N^T N$ is the positive-definite communication-line impedance matrix, $\dot{q}_1^d$ is the desired velocity for the master system, τ_1 will embed the PI-control (with respect to $\dot{q}_1 - \dot{q}_1^d$) to relate $\dot{q}_1$ and $\dot{q}_1^d$, and s_*^+, s_*^- are the outgoing and incoming scattering variables. See Figure 3 with the indexes h, L replaced by 1, 2. The communication law is then given by

$$s_1^-(t) = s_2^+(t - \lambda_{21}), \ \ s_2^-(t) = s_1^+(t - \lambda_{12}),$$

where λ_{12} is the delay from the master to the slave. For the master scattering block, the inputs are (s_1^-, τ_1), while the outputs are $(s_1^+, \dot{q}_1^d)$. As shown in Niemeyer and Slotine (2004), this scattering technique can enforce two-port closed-loop passivity of the telerobotic system even with varying delay and data loss. Yet, due to the lack of explicit

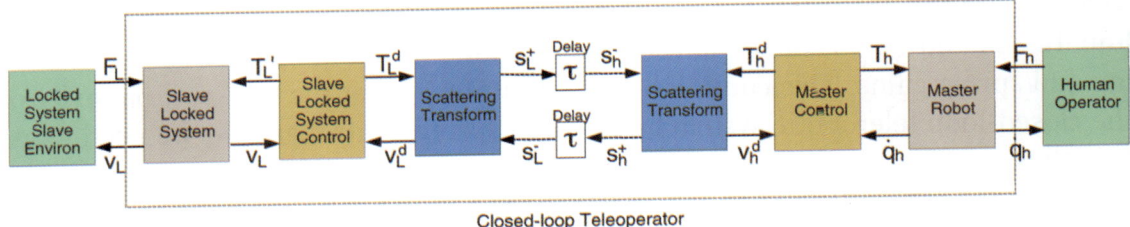

Figure 3. Scattering-based teleoperation.

position feedback (i.e., the scattering variable contains information only on velocity, not position), a bit of data loss may cause a position-drift problem (Niemeyer & Slotine, 1998). Also, the wave-reflection phenomenon, which often produces sustained oscillations in the closed-loop telerobotic system, may occur, whose remedy seems only applicable to linear telerobotic systems (Niemeyer & Slotine, 2004).

Recently, it was shown by Lee and Spong (2006) that the simple PD control (Equation 4) can also enforce the two-port passivity of telerobotic systems with constant communication delays. If we use a control such as

$$\tau_1 = -B_e \dot{q}_1 - B(\dot{q}_1 - \dot{q}_2(t - \lambda_{21})) - K(q_1 - q_2(t - \lambda_{21}))$$

for the master side and a similar control for the slave side, we can tune the gains such that the following condition is satisfied—

$$B_e \succ \frac{\max(\lambda_{12} + \lambda_{21})}{2} K$$

(where $A \prec B$ means $A - B$ is positive-definite), the closed-loop telerobotic system will be two-port passive. This condition states that the dissipation gain B_e needs to be large if we want a large K (i.e., high performance) and the round-trip delay $\lambda_{12} + \lambda_{21}$ is large. The damping term B in the above τ_1 can be omitted without compromising the closed-loop passivity. This result is the very first passivity-enforcing non-wave-based telerobotic control law with constant time delay, which has been ensued by some theoretical refinements (Kawada & Namerikawa, 2004; Nuno, Ortega, Barabanov, & Basanez, 2008). We believe that this PD-based technique, due to its simplicity, will be useful for many virtual and remote manufacturing applications, particularly their rapid prototyping.

If the round-trip delay $\lambda_{12} + \lambda_{21}$ gets longer (e.g., longer than 1 sec. [Lee & Spong, 2006]), the required damping B_e will also need to be larger, making the resultant system exhibit sluggish behavior with substantial reaction force in the free-space motion due to the large damping B_e. This PD technique and its passivity argument are also applicable only to constant time delay, although some experiments show its efficacy for an Internet-like communication scenario (Rodriguez-Seda, Lee, & Spong, 2006). An extension of this PD control to the Internet-like communication—passive set-position modulation (PSPM)—is proposed by Lee and Huang (2008b), where the following PD-type control (for the master—similar also for the slave) is used:

$$\tau_1 = -B\dot{q}_1(t) - K(q_1(t) - y(k)),$$

where $y(k)$ is the desired set position received from the Internet-like communication, which is given by a switching sequence of discrete-signals, with varying-delay/packet loss and even time swapping possibly embedded in this sequence (see Figure 4a, with x replaced by q_1). As shown in Lee and Huang (2008b), the only reason the telerobotic system is not two-port passive is the spring potential energy jumps at each switching instance:

$$\Delta P(k) := \varphi(t_k) - \varphi(t_k^-) \neq 0,$$

where $\varphi(t) := \frac{1}{2}(q_1(t) - y(k))^T K(q_1(t) - y(k))$ for $t \in I_k := (t_k, t_{k+1})$. If there is a certain sequence of $y(k)$ that induces a series of positive energy jumps so that $\sum \Delta P(k) \to +\infty$, the energy in the closed-loop telerobotic system will keep increasing, violating the two-port closed-loop passivity. Note here that there is no guarantee that such a passivity-breaking sequence does not exist, since the Internet communication is itself random.

To address this problem, the PSPM technique (Lee & Huang, 2008b) observes such spring energy jumps and regulates them so that the closed-loop system can still be two-port passive. At each switching instance t_k, we solve the following optimization problem:

$$\min_{\bar{y}(k)} \quad \left\| y(k) - \bar{y}(k) \right\|$$

$$\text{subj. } E(k) = E(k-1) + \Delta E_y(k) + D_{\min}(k-1) - \Delta \bar{P}(k) \geq 0,$$

where $\bar{y}(k)$ is the modulated version of $y(k)$, which allows us to avoid a passivity-breaking energy jump and $\Delta \bar{P}(k)$ is the spring energy jump when $\bar{y}(k)$ is used instead of $y(k)$. The term $E(k)$ represents a virtual energy storage (simulated in PC), and

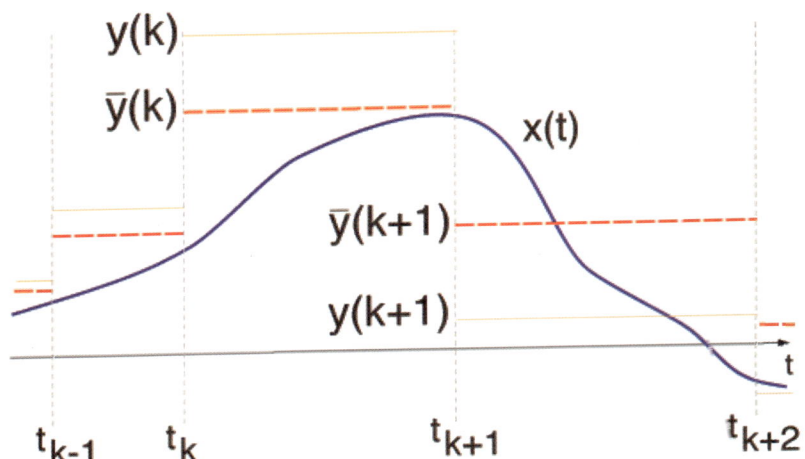

(a) Discrete $y(k)$

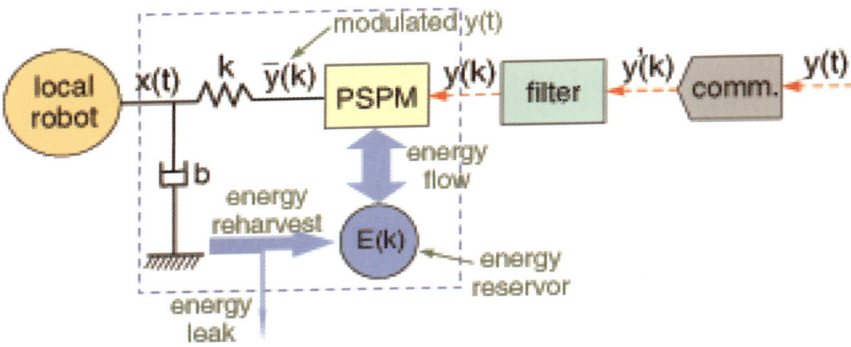

(b) Passive set-position modulation (PSPM)

Figure 4. Discrete update of set position $y(k)$ and PSPM architecture (from Lee & Huang, 2008b).

$\Delta E_y(k) > 0$ is an energy-shuffling term sent from the slave side. This $E(k)$ also recaptures (otherwise wasted) energy dissipation through the control damping B via the term

$$D_{\min}(k) := \frac{1}{t_{k+1}-t_k}\sum_{i=1}^{N} b_{ii}(\bar{q}_1^i(k)-\underline{q}_1^i(k))^2 \geq \int_{t_k}^{t_{k+1}} \left\| \dot{q}_1 \right\|_B^2 dt,$$

where b_{ii} is the diagonal element of the diagonal control damping B and $\bar{q}_1^i(k), \underline{q}_1^i(k)$ are the maximum and minimum of the ith component of q_1 during (t_k, t_{k+1}), which can usually be measured by encoders. See Figure 4b for an illustration of this architecture, where the energy shuffling is combined into the energy flow between the PSPM block and $E(k)$.

Note that this PSPM aims to choose $\bar{y}(k)$ as close to $y(k)$ as possible, but only to the extent permissible by the available energy in the system. As shown in Lee and Huang (2008b), by using exactly the same PSPM on both the master and slave sides, the closed-loop telerobotic system will be guaranteed to be two-port passive. As we only require the incoming data stream $y(k)$ to be a discrete sequence, this PSPM can also be applied for a variety of communication imperfections such as varying delay, packet loss, and even time swapping. For the same reason, some discrete data processing can be also inserted in the "filter" block in Figure 4b while enforcing the two-port passivity of the closed-loop telerobotic system. As the passivity is enforced not only by a single damping gain B (which needs to be activated all the time), but also by an additional PSPM algorithm, the required damping level B is much lower than that of the PD-control case, allowing us to avoid a sluggish system response of the PD control with long delays. This PSPM is the very first theoretical passivity-enforcing result for telerobotic systems with the Internet-like communication with transmission of explicit position information. For further details, see Lee and Huang (2008b).

5   Conclusions

In this chapter, we introduced certain ideas and concepts of telerobotics that may be useful for some applications in virtual and remote manufacturing applications. In particular, we explained the notions of performance (i.e., ideal transparency) and stability (i.e., passivity). We also presented approaches to achieve motion- and power-scaling (Section 2) and virtual constraints (Section 3). In addition, we showed how to connect the master and slave systems while enforcing passivity when the master-slave communication is potentially imperfect. The concepts presented in this chapter also apply to telerobotics (Hwang & Hashimoto, 2007; Lee & Spong, 2005) and teleoperation of wheeled mobile robots, which are becoming more common in manufacturing (Diolaiti & Melchiorri, 2003; Lee, 2008; Lee, Martinez-Palafox, & Spong, 2006; Lee, Sukhatme, Kim, & Park, 2002).

References

Anderson, R. J., & Spong, M. W. (1989). Bilateral control of tele-operators with time delay. *IEEE Transactions on Automatic Control, 34*(5), 494–501.

Barbič, J., & James, D. L. (2008). Six-DoF haptic rendering of contact between geometrically complex reduced deformable models. *IEEE Transactions on Haptics, 1*(1), 1–14.

Bettini, A., Marayong, P., Lang, S., Okamura, A. M., & Hager, G. D. (2004). Vision-assisted control for manipulation using virtual fixtures. *IEEE Transactions on Robotics, 20*(6), 953–966.

Brown, J. M., & Colgate, J. E. (1998). Minimum mass for haptic display simulations. *Proceedings of ASME International Mechanical Engineering Congress and Exposition,* 249–256.

Colgate, J. E., & Schenkel, G. (1997). Passivity of a class of sampled-data systems: Application to haptic interfaces. *Journal of Robotic Systems, 14*(1), 37–47.

Diolaiti, N., & Melchiorri, C. (2003). Haptic tele-operation of a mobile robot. *Proceedings of the 7th IFAC Symposium of Robot Control,* 2798–2805.

Duindam, V., & Stramigioli, S. (2004). Port-based asymptotic curve tracking for mechanical systems. *European Journal of Control, 10*(5), 411–420.

Goldfarb, M. (1999). Similarity and invariance in scaled bilateral telemanipulation. *ASME Journal of Dynamic Systems, Measurements, and Control, 121*(1), 79–87.

Hannaford, B. (1989). A design framework for teleoperators with kinesthetic feedback. *IEEE Transactions on Robotics and Automation, 5*(4), 426–434.

Hannaford, B., & Ryu, J.-H. (2002). Time domain passivity control of haptic interfaces. *IEEE Transactions on Robotics and Automation, 18*(1), 1–10.

Hogan, N. (1989). Controlling impedance at the man/machine interface. *Proceedings of the 1989 IEEE International Conference on Robotics and Automation,* 1626–1631.

Hokayem, P. F., & Spong, M. W. (2006). Bilateral teleoperation: An historical survey. *Automatica, 42,* 2035–2057.

Hwang, G., & Hashimoto, H. (2007). Development of a human-robot-shared controlled teletweezing system. *IEEE Transactions on Control Systems Technology, 15*(5), 960–966.

Kawada, H., & Namerikawa, T. (2004). Bilateral control of nonlinear teleoperation with time varying communication delays. *Proceedings of the American Control Conference,* 189–194.

Khatib, O. (1986). Real-time obstacle avoidance for manipulators and mobile robots. *International Journal of Robotics Research, 5*(1), 90–98.

Koditschek, D. E. (1991). The control of natural motion in mechanical systems. *ASME Journal of Dynamic Systems, Measurements, and Control, 113,* 547–551.

Kosuge, K., Itoh, T., & Fukuda, T. (2000). Human-machine cooperative telemanipulation with motion and force scaling using task-oriented virtual tool dynamics. *IEEE Transactions on Robotics and Automation, 16*(5), 505–516.

Lawrence, D. A. (1993). Stability and transparency in bilateral teleoperation. *IEEE Transactions on Robotics and Automation, 9*(5), 624–637.

Lee, D. J. (2008). Semi-autonomous teleoperation of multiple wheeled mobile robots over the Internet. *Proceedings of ASME Dynamic Systems & Control Conference.* Also available at http://web.utk .edu/~djlee/papers/DSCC08a.pdf

Lee, D. J., & Huang, K. (2008). On passive non-iterative variable-step numerical integration of mechanical systems for haptic rendering. *Proceedings of ASME Dynamic Systems & Control Conference.* Also available at http://web.utk.edu/~djlee/papers/DSCC08b.pdf

Lee, D. J., & Huang, K. (2008). Passive position feedback over packet-switching communication network with varying-delay and packet-loss. *Proceedings of Symposium of Haptic Interfaces for Virtual Environments & Teleoperator Systems,* 335–342.

Lee, D. J., & Li, P. Y. (2003). Passive bilateral feedforward control of linear dynamically similar teleoperated manipulators. *IEEE Transactions on Robotics and Automation, 19*(3), 443–456.

Lee, D. J., & Li, P. Y. (2005). Passive bilateral control and tool dynamics rendering for nonlinear mechanical teleoperators. *IEEE Transactions on Robotics, 21*(5), 936–951.

Lee, D. J., Martinez-Palafox, O., & Spong, M. W. (2006). Bilateral teleoperation of a wheeled mobile robot over delayed communication networks. *Proceedings of IEEE International Conference on Robotics and Automation,* 3298–3303.

Lee, D. J., & Spong, M. W. (2005). Bilateral teleoperation of multiple cooperative robots over delayed communication networks: Theory. *Proceedings of IEEE International Conference on Robotics and Automation,* 362–367.

Lee, D. J., & Spong, M. W. (2006). Passive bilateral teleoperation with constant time delay. *IEEE Transactions on Robotics, 22*(2), 269–281.

Lee, S., Sukhatme, G., Kim, G. J., & Park, C. (2002). Haptic teleoperation of a mobile robot: A user study. *Presence, 10*(11), 1309–1313.

Li, P. Y., & Horowitz, R. (1999). Passive velocity field control of mechanical manipulators. *IEEE Transactions on Robotics and Automation, 15*(4), 751–763.

Liberzon, D., & Morse, A. S. (1999). Basic problems in stability and design of switched systems. *IEEE Control Systems Magazine, 19*(5), 59–70.

Niemeyer, G., & Slotine, J. J. E. (1991). Stable adaptive teleoperation. *IEEE Journal of Oceanic Engineering, 16*(1), 152–162.

Niemeyer, G., & Slotine, J. J. E. (1998). Towards force-reflecting teleoperation over the Internet. *Proceedings of IEEE International Conference on Robotics and Automation,* 1909–1915.

Niemeyer, G., & Slotine, J. J. E. (2004). Telemanipulation with time delays. *International Journal of Robotics Research, 23*(9), 873–890.

Nuno, E., Ortega, R., Barabanov, N. E., & Basanez, L. (2008). A globally stable PD-controller for bilateral teleoperators. *IEEE Transactions on Robotics, 24*(3), 753–758.

Oboe, R., & Fiorini, P. (1998). A design and control environment for Internet-based telerobotics. *International Journal of Robotics Research, 17*(4), 433–449.

Peshkin, M., & Colgate, J. E. (1999). Cobots. *Industrial Robot, 26*(5), 335–341.

Rodriguez-Seda, E. J., Lee, D. J., & Spong, M. W. (2006). An experimental comparison of bilateral Internet-based teleoperation. *Proceedings of CCA/CACSD/ISIC.*

Rosenberg, L. (1993). Virtual fixtures: Perceptual tools for telerobotic manipulation. *Proceedings of the IEEE Virtual Reality International Symposium,* 76–82.

Stramigioli, S., van der Schaft, A., Maschke, B., & Melchiorri, C. (2002). Geometric scattering in robotic telemanipulation. *IEEE Transactions on Robotics and Automation, 18*(4), 588–596.

Vidyasagar, M. (1993). *Analysis of nonlinear dynamic systems.* 2nd ed. Englewood Cliffs, NJ: Prentice Hall.

Yokokohji, Y., & Yoshikawa, T. (1994). Bilateral control of master-slave manipulators for ideal kinesthetic coupling—formulation and experiment. *IEEE Transactions on Robotics and Automation, 10*(5), 605–620.

8 | Barriers to Collaboration in Global Teams

A Case Study

Caroline Clarke Hayes, Ph.D.

Department of Mechanical Engineering, University of Minnesota

Abstract

Geographically distributed product development teams have become a fact of life for international companies. This chapter presents a case study of a company that moved their manufacturing facility from a location next door to the design offices to a distant location where skilled labor was less expensive. Over time, they found that product yields were very low for some new products (but not others), which threatened the company's ability to release new products and stay competitive. Managers described that the problem was a "lack of communication" between sites; however, they could not pinpoint the major causes, nor could they identify lasting solutions. The first step toward solutions is the identification of underlying problems and mechanisms. The case study examines both sites in a series of interviews with employees and managers and describes some of the major collaboration barriers between sites, including the role of "ad hoc consultants." While the majority of collaboration barriers are described in other places in the abstract, the value of a case study is to illustrate how these barriers present themselves in a specific company and situation. While distance collaboration problems may be common to many organizations, the degree of those problems, the impact on the business, and the underlying causes (and hence the solutions) may arise from that organization's specific work processes, management structures, priorities, and incentive systems. This case study provides concrete examples that may be useful to others in uncovering and characterizing their own collaboration challenges.

Keywords: collaboration barriers, distance collaboration, global teamwork

1 Introduction

Virtual and global teams have become common and necessary for many companies in the 21st century. *Virtual teams* are those that communicate primarily through electronic media (Kunz, Christiansen, Cohen, Jin, & Levitt, 1998); a *global team* is one in which team members conduct their work from multiple locations, possibly in different countries, and in which members have diverse cultural backgrounds. Global design teams often carry out many (but not all) of their interactions virtually because of the cost and time involved in flying all team members to one location.

While global teams are not new, their recent increase in popularity is largely due to widespread availability of a variety of high-quality, low-cost communication media including e-mail, Internet, faxes, and teleconferencing. These media have made national and international communication and data exchange possible in ways that were not feasible 20 years ago and have hugely reduced the cost of conducting global teamwork. Global design teams can provide many benefits; for example, they allow companies to take advantage of a broader range of expertise than is available at one site or to make use of skilled manufacturing labor and expertise in locations where skilled labor is less expensive.

However, global teams are often less effective than desired (McDonough, Kahnb, & Barczaka, 2003). Communication and collaboration challenges responsible for a lack of effectiveness can often be subtle and maddeningly difficult to characterize, yet they can have profound impacts on product development time, employee satisfaction, product cost and quality, and ultimately, company viability.

This chapter (a) summarizes literature on the benefits and challenges of virtual and global teams, with a particular focus on product design teams, and (b) presents a case study of cross-site collaboration challenges observed in a biomedical product company.

The case study illustrates not only how the collaboration challenges that are abstractly described in the literature manifest themselves in an actual product development team but also how the subtle, complex, interactive, and multifaceted nature of collaboration challenges.

2 The Challenge

2.1 Company Characteristics

The company in the case study was a designer and manufacturer of implantable medical devices that are surgically placed in the body and help to regulate the heart or other organs by delivering electrical impulses or medicines. Surgeons will choose those products that not only are reliable and safe but also incorporate the latest medical advances to make them maximally effective.

Success in such a business is highly dependent on continual research and new product development to continually broaden the functions and improve the efficacy, safety, and comfort of the products. Thus, product runs tend to be relatively short and small in volume (e.g., when compared to the manufacture of cell phones); any given product may be produced for only a short time before a new and improved version comes out. Thus, it is important that there are few "bugs" in the manufacturing process. Delays in production to work out "bugs" can result in a missed window of opportunity for that product; critically ill patients cannot benefit from a new technology that is not produced, or market shares may be lost if a competitor comes out with a similar technology first.

2.2 The Situation History

Several years prior to the study, the company's design offices and manufacturing facility were located across a parking lot from each other. The manufacturing facility was relocated to a place where skilled labor was less expensive (Figure 1). The preferred language at the design offices was English, and the preferred language at the manufacturing site was Spanish. Most manufacturing engineers and managers were also fluent in English, but the assemblers were not.

Over time, long after initial training and operation setup were completed in the new manufacturing facility, they found that product yields were very low for some new product lines (but not others). Product yield is the percentage of "good" verses "scrap" products produced per time unit.

This threatened the company's ability to release new products and therefore to stay competitive and thrive. Managers characterized the problem as a "lack of communication" between sites; however, they could not pinpoint the mechanisms

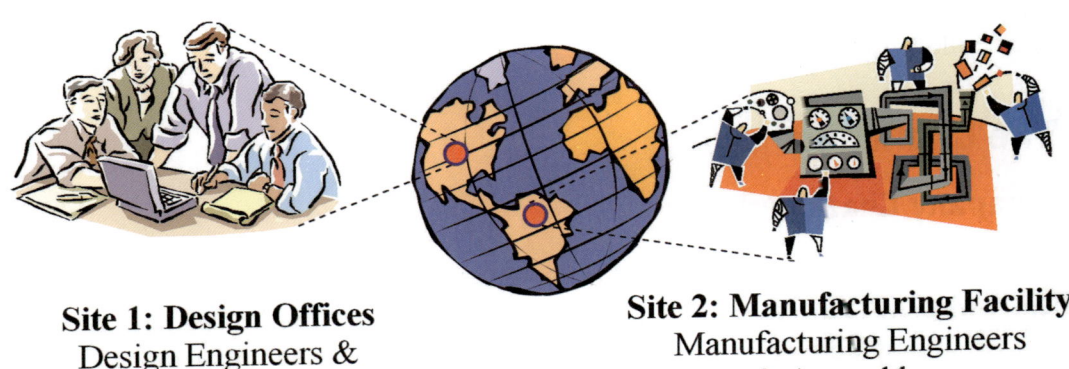

Site 1: Design Offices
Design Engineers &
Medical Experts

Site 2: Manufacturing Facility
Manufacturing Engineers
& Assemblers

Figure 1. The design and manufacturing offices were separated by great distances.*

* This figure does not represent the actual locations of the facilities.

resulting in a lack of communication, nor could they identify lasting solutions that were more concrete than "people should talk more across sites."

The company commissioned a team of engineering researchers, including the author, to study the situation. The goals of the study were to identify (a) the nature of the problem (more specifically than "people don't talk enough across sites"), (b) the mechanisms and causes underlying the problems, and (c) to recommend solutions where possible.

3 Approach

This study was exploratory in nature, aimed at framing the problem and identifying systemic issues that existed between design and manufacturing groups within the company. Data and information were gathered over a period of 2 days at the design offices and 4 days at the manufacturing facility through a combination of techniques, including the following (Wickens, Lee, Liu, & Becker, 2004):

- Focus groups composed of managers, design engineers, and manufacturing engineers
- Structured and open-ended interviews with representatives from these groups
- Presentations by members of these groups describing their design processes and their views of the problems

- Historical data on the type and frequency of specific types of manufacturing defects

- Site "walk-throughs" of the manufacturing facility to observe manufacturing processes in action

The following section describes some of properties of global and virtual teams identified in the literature. Much of this literature focuses on interpersonal relationships and communication between recognized members of the design team. While the case study found all these issues to be important, its observations additionally highlighted (a) the critical role that the organization plays in supporting (or not supporting) interactions across distance through the priorities, resources, and incentives provided to team members and (b) the critical role that people who are not explicitly recognized as part of the product development team, such as ad hoc consultants, can play in the team's success.

4 Related Literature

4.1 The Increasing Prevalence of Virtual and Global Teams

McDonough, Kahnb, and Barczak (2001) found that more than 50% of the 103 new product development companies they surveyed used global teams for product development. For the companies using global teams, approximately one in five teams were likely to be global. Use of global teams has continued to increase since that time as companies have become more globally dispersed.

4.2 Challenges of Virtual and Global Teams

Global teams often include more diverse expertise, perspectives, and cultures than colocated teams. However, this is both their strength and their weakness. McDonough, Kahnb, and Barczak (2001) found global team performance to be lower than that of either colocated or (nonglobal) virtual teams. Similarly, Distefano and Maznevski (2000) found that diverse, multicultural teams (terms that aptly describes most global design teams) perform either worse or better than homogeneous teams, with worse performance being more common. This is in part because of the greater knowledge, communication, negotiation, and diplomacy skills needed when working with people from other disciplines, organizations, and cultures.

The nature of global design teams not only exacerbates the problems of traditional cross-functional teams but also creates new ones (McDonough, Kahn, & Griffin, 1999). Consequently, they are more difficult to manage. In a survey of 109 companies,

Barczak and McDonough (2003) categorized these challenges as interpersonal and programmatic:

- *Interpersonal issues* included finding ways to build working relationships and trust between distant team members and keeping team morale high in the absence of frequent face-to-face meetings.

- *Programmatic issues* included keeping the project on schedule, procuring adequate resources, and keeping team members focused on the project, despite competing pressures from local site managers and the "out-of-sight, out-of-mind" phenomena.

Unfortunately, the understanding of how to manage global design teams has not kept pace with their increasing use (Barczak & McDonough, 2003). While much of the literature focuses on interpersonal issues, the case study described in this chapter found the programmatic issues to dominate the interpersonal issues. The description of these issues can help to increase understanding of programmatic issues, which may in turn facilitate development of appropriate global management strategies.

4.3 Advantages of Face-to-Face and Colocated Teams

There is much work focused on explaining why *colocated* (i.e., same site but not necessarily face-to-face) and face-to-face (same room) design teams may function more effectively, on average, than global teams:

- *Communication effort.* The bar for communication is lower in colocated teams because it takes less effort to set up meetings and there are more opportunities for face-to-face meetings.

- *Media richness.* Face-to-face communication is "richer" than most of the electronic communication that comprises much of the interaction between distant design team members. It is richer in that it provides multiple channels for communication in addition to verbal inflection, body language, and gestures, and it provides more opportunities for interactive feedback (Daft & Lengel, 1986).

- *Building interpersonal relationships and trust.* This richness is hypothesized to be important in supporting the exchange of ambiguous information (Song, Berends, van der Bij, & Weggeman, 2007), such as partially formulated design concepts, and in building interpersonal relationships, which are critical to the functioning of any team. In particular, Handy (1995) found that face-to-face meetings are important to building trust between participants.

- *Coordination latencies.* The type of design process followed can also have a great impact on design team performance in terms of project duration and cost. Chachere, Kunz, and Levitt (2004) use the concept of "coordination latency" to explain why design project durations are dramatically reduced from months to days when design teams follow a design process called *extreme collaboration.* In extreme collaboration, design team members not only are located at the same site but also carry out much of their design work in the same room and at the same time in scheduled design sessions. When designers work, questions concerning related subsystems frequently arise. When a question arises during design work on another subsystem, if most of the design team is located in the same room, the designer can quickly look around the room for team members who can answer her question. She can often get the answer in seconds to minutes.

In contrast, when designers work in separate offices (possibly at the same site), they must call or e-mail one another to get questions answered. If the respondent is not in or is engaged in other projects, it can take hours to weeks to get questions answered. This occurs many, many times during a typical design project. When an extreme collaboration process is employed, the cumulative reduction of many small delays can result in immense reductions in the overall project duration. In contrast, members of a globally distributed design team typically conduct most of their project work asynchronously from their separate offices. Coordination latencies can be very large.

When teams become virtual or virtual and geographically distributed, they lose many of the advantages enjoyed by face-to-face colocated teams, while the challenges associated with multiple time zones and cross-cultural interactions are added (McDonough, Kahn, & Griffin, 1999).

5   The Case Study Findings

Interviews and observations of the design and manufacturing sites clarified the way in which product design teams in this industry were composed and operated. The study also revealed an overwhelming and diverse range of collaboration issues and challenges for virtual and global teams. The first challenge was to categorize and then make sense of these issues.

While the literature on virtual and global teams makes clear *what* general types of collaboration challenges and problems exist for such teams, the case study illustrates *how* these problems arose when product design teams became global. The following are some general themes that emerged:

- The effort and resources required for successful global team collaboration was often greater than for colocated teams. However, additional resources and incentives for additional effort were not provided, nor was the need recognized.

- Informal processes that make colocated teams "work" were often not recognized as important. Thus, when distance makes these informal processes difficult or infeasible, it may be difficult to identify the lack of these "invisible" informal processes as the cause of problems. An example is the informal understanding of which people have what expertise in another group. Distance reduces this understanding.

- People who are not formally part of a product design team are often important contributors. Distance made it more difficult to find and consult these people at the other site. However, the informal nature of these "ad hoc consultants" reduced their visibility, making their lack difficult to identify as a "problem."

Finally, multiple factors typically contributed to any individual collaboration problem, making it difficult to understand causes. A few of these findings are presented below.

5.1 Special Properties of the Design Process

A critical aspect of all medical product development is that designs must pass through a lengthy federal Food and Drug Administration (FDA) approval process. This process requires many animal and human clinical trials with product prototypes and can take 2 years or more to complete. Once a design passes the FDA approval process, it is considered to be "frozen"; design changes after this point are usually infeasible due to the great time and expense involved in additional clinical trials required for FDA approval of changes. This made it particularly important to identify and resolve any manufacturing issues that could be solved through design changes *before* starting the FDA approval process. The company knew this and therefore explicitly included manufacturing reviews at several points in the formal documented design process applied to all products. Additionally, the formal process called for manufacturing test runs of prototypes before entering the FDA approval process. However, for reasons that will be described later, these design reviews and manufacturing test runs were often given short shrift.

5.2 Design Team Composition

Not everyone on a product design team is necessarily a "designer" or an engineer. Successful design of medical products requires integration of a wide range of expertise in diverse areas such as medicine, mechanical systems, electrical systems, materials, and manufacturing methods, as shown in Figure 2. This approach to design is often called *design for life cycle* (Ishii, 1995), in which representatives from many aspects of the product's life cycle participate in the design process. In Figure 2, most of the people shown at the design offices (Site 1) were design engineers and medical experts, although a few were manufacturing engineers. The people at the manufacturing facility (Site 2) included both college-educated manufacturing engineers and highly skilled assemblers who carried out the actual manufacturing and who were not typically college educated.

While design may not be the primary training or responsibility of most manufacturing engineers, it is important for them to be involved at multiple points in the design process so that they may provide input regarding manufacturing feasibility and cost for the options under consideration, identify manufacturing issues associated with the current design, and jointly brainstorm with design engineers on design modifications that can lower manufacturing costs without compromising the functions of the device. The practice of simultaneously designing the product and the

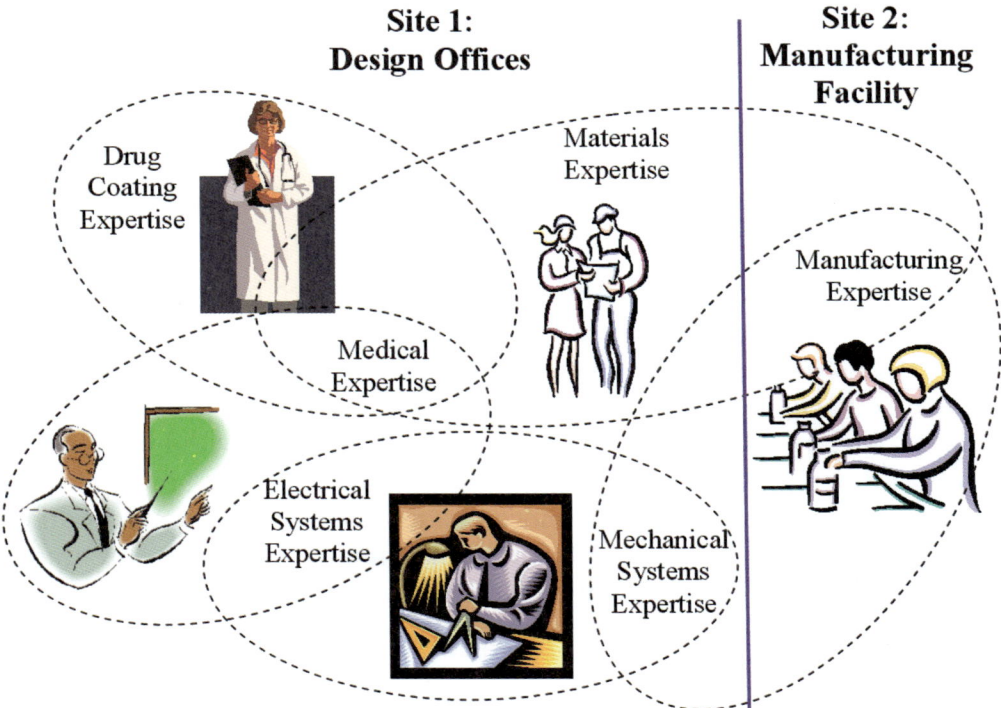

Figure 2. Design team members have overlapping expertise.

manufacturing process by which it will be made is referred to as *concurrent product/ process engineering* (Sohlenius, 1992), or simply *concurrent engineering*.

When manufacturing engineers do *not* participate in the design process, the resulting designs are less likely to be easy to manufacture. In the absence of expert manufacturing advice, designers must use their own less informed and less accurate understanding of manufacturing to choose between design options. Sometimes they did not make appropriate choices, which gave rise to many of the low product yields experienced by this company.

5.3 The Role of Ad Hoc Consultants

Design team members may be formally assigned to a design project, or they may be informally consulted on an as-needed basis. Manufacturing engineers were often informal members of the design team, contacted on an ad hoc basis to resolve specific questions as they arose. For example, if a designer wanted to join two sections of tubing in a way that had not been tried previously in other products, she might know of a particular person in manufacturing with special experience in material welding whom she could call to discuss the feasibility.

However, when design and manufacturing offices were moved to a distant site, designers were far less likely to know the manufacturing engineers or their capabilities. If a designer did not know whom to call to discuss a manufacturing issue, the issue often went unresolved or was solved suboptimally using the design engineers' more limited understanding of the manufacturing process.

6   Discussion

The design-manufacturing collaboration barriers identified in the study were categorized as either interpersonal or systemic, although there is not necessarily a distinct line between the two. Barriers categorized as *interpersonal* were those that had a fairly direct impact on peoples' abilities to communicate, collaborate, or form relationships across sites. Barriers categorized as *systemic* were associated with high-level process and organizational issues and often had a less direct but no less powerful impact on collaboration across sites. We have chosen to use the term "systemic" rather than Barczak and McDonough's (2003) term "programmatic" because "programmatic" suggests issues relating to individual projects and programs, while many of the issues observed in this study were problems that were repeated in many projects and pervaded the whole organization.

Following are examples of specific barriers to collaboration that existed in the company, as identified by the study. They will serve as a basis for identifying parameters that impacted cross-site collaboration. They also provided insights into changes to recommend for improvement.

6.1 Interpersonal Barriers

Like the study reported by McDonough, Kahn, and Griffin (1999), this study found that distance exacerbated existing problems experienced by colocated cross-functional (design and manufacturing) teams and added new ones. The historical problems faced by colocated design and manufacturing groups are long standing and well documented. They include differences in culture (design has a white-collar culture, while manufacturing typically has a blue-collar culture), work environment (design occurs in an office setting and manufacturing in factories), and power (design is usually considered to be of a higher stature than manufacturing), which are reinforced through differences in education and salary. These differences at best result in less than optimal interactions when design and manufacturing are colocated, resulting in the proverbial "wall" between design and manufacturing. The added challenges of distance collaboration previously described further exacerbated these traditional challenges in the following ways:

1. *Lost familiarity with people and their skills across sites.* While moving the manufacturing facility to a location with lower labor costs solved one problem for the company, it also introduced some new ones that it did not anticipate. In particular, people at the design offices lost familiarity with the people at the manufacturing site and vice versa. This included a loss of awareness of who was an expert in a particular subject area at the other site, making it difficult to know whom to consult at the other site.

2. *Lost cross-site domain knowledge.* Additionally, when design and manufacturing groups were colocated, people more readily acquired some degree of "overlapping knowledge" and capability in the domain of expertise of the other group. For example, many designers acquired fairly detailed manufacturing knowledge. This gave them common ground and vocabulary that enabled them to more readily discuss manufacturing issues with the manufacturing engineers. This acquisition of knowledge from those with whom one frequently interacts resulted in overlapping areas of expertise, as shown in Figure 2. The same was true for manufacturing engineers and assemblers; while the assemblers were not formally trained as designers, many of them acquired fairly detailed design knowledge. However, when design and manufacturing offices were moved to distant locations, it became much more difficult for designers to acquire and maintain current manufacturing knowledge, and likewise for manufacturing engineers. This contributed to the next issue described.

3. *Lack of common training and common ground.* The designers did have a basic understanding of what the manufacturers did and vice versa. However, that understanding was often not sufficiently deep to allow them to make effective suggestions to each other. For example, a manufacturing engineer reported asking a designer, "Can we change feature x to reduce the manufacturing cost?" The designer responded with a long, complicated medical explanation of what the feature did when placed inside the human body and why it could not be changed in that way. The manufacturer, lacking the medical background necessary to understand the explanation, simply respond with "OK," rather than attempting to suggest another alternative that would accomplish the same purpose at a lower cost. Additionally, a limited understanding of the functional issues limits manufacturers' ability to make viable redesign suggestions. Similarly, although when designers were first employed, they were given some training at the manufacturing site on the companies specialized manufacturing processes, the manufacturing processes were continually changing and improving. Thus, the designers' manufacturing knowledge rapidly became out of date. Thus, their ability to independently create manufacturable designs decayed over time.

4. *Power differences exacerbated between sites.* Company personnel did not specifically report cross-cultural issues as a communication barrier; however, the culture differences likely exacerbated the power differences between the sites, and power differences were reported as an issue that impacted communication. Dominance impacted communication in that participants from the nondominant site reported that they were less likely to initiate communication with the dominant site, and their input was also less likely to be heard, which led to the next issue.

5. *Manufacturing voices not heard in design reviews.* Because of the power differences previously described, the input of manufacturing engineers sometimes was not given sufficient weight in design decisions, just as nurses' input is often not "heard" by doctors. This was true when design and manufacturing were colocated, but it became even easier to ignore manufacturing voices after they moved.

The combined impact on collaboration between design and manufacturing offices was devastating. Designers and manufacturers were far less likely to know whom to call at the other site if a problem occurred. If they did figure out whom to call, they had less common ground to enable them to understand and negotiate issues. And if they could not figure out whom to call, they possessed limited cross-site expertise to enable them to develop a good solution on their own.

6.2 Systemic Barriers

Systemic barriers were often subtle, complex, and indirect in their effects on collaboration and on product yields. Barriers in this category are typically associated with high-level organization-wide issues, including management approaches, formal design processes followed throughout the organization, and assignment of priorities, resources, responsibilities, and rewards to people in various roles.

For example, designers' primary responsibility and priority was to get designs completed and "out the door" on schedule so the FDA approval process could begin and new design projects could be started. Designers were not responsible for what happened in manufacturing when the design finally reached the production stage 1 or 2 years later. Consequently, their motivation to negotiate and resolve manufacturing issues was not their top priority. This is an example of a systemic barrier to collaboration resulting from unintended consequences of the organization's priorities, accountability, and reward system. Several examples of systemic barriers identified in the study are as follows:

1. *Limited time for new manufacturing process development and testing.*
   Designers reported that some design features inherently led to high scrap rates. For example, a designer might choose to coat a portion of the device with a material that is both tolerated by human tissues and has the required drug permeability, despite that they know this material stains easily during the handling required for assembly. Stained products must be scrapped. However, designers choose to include these materials in new designs knowing it may reduce yields because there are no alternative FDA-tested materials that can reliably meet the functional requirements.

2. *Unresolved manufacturing issues.* Sometimes designers identified manufacturing issues during the design's manufacturing verification or prototyping *prior* to the start of the FDA approval process. Some of those issues could be resolved through design modifications, but this did not always happen because of time pressures to "get the design out the door." One designer stated that "manufacturing can work it out later," which they usually could, but often with great cost, reduced yields, and high personal stress.

3. *Lost manufacturing wisdom at the end of product cycles.* There are many process improvements and innovations that may occur during the manufacturing run of a specific product. Manufacturing personnel can learn many valuable lessons that could be used profitably in future product designs. However, these lessons were often lost because there was no formal mechanism for recording them.

7   Summary

What follows is a "nutshell" summary of why the design and manufacturing groups observed failed to collaborate sufficiently. The physical distance between design and manufacturing increased the effort required to communicate and collaborate across sites; individual communication was less frequent and of lower quality. This had two primary effects: people at each site had less "expertise overlap" with people at the other site, and they had less awareness of who possessed what expertise at the other site. The first made it more difficult for people from the two sites to discuss issues in a common technical language or to even understand each other's problems. The second made it more difficult for designers to identify manufacturing consultants when specific manufacturing questions arose. Thus, manufacturing issues were often left unresolved at design completion. Overall, communication and collaboration between the sites became more time consuming, frustrating, and costly. However, management had not changed the rewards, incentives, or resources available to motivate people to make the extra effort required. Thus, cross-site collaboration suffered and was manifested as poor product yields.

8   Recommendations

Recommendations focused on the following issues:

- *Lost cross-site expertise overlap between sites.* This included the design engineers' loss of familiarity with the current (continually evolving) manufacturing techniques and loss of manufacturing wisdom and "lessons learned" at the end of a product run. Thus, designers did not get sufficient feedback on design choices that created manufacturing difficulties and thus kept making the same mistakes.

- *Lost awareness of who had expertise in what area at the other site.* This made it difficult for designers find manufacturing consultants to help them make design choices that minimize manufacturing difficulties. It also made it difficult for manufacturers to identify design consultants to help them identify approaches for resolving manufacturing problems that would not compromise the product's intended functions.

Recommendations that addressed one or both of these issues simultaneously included the following:

- *Formation of a special "manufacturing knowledge-capture" team* composed of design and manufacturing engineers from both sites dedicated to collecting, recording, and disseminating "lessons learned" from each manufacturing run. This serves to address both the aforementioned issues. Design and manufacturing engineers from different sites who are brought together by participation on this team get to know each other and develop overlapping expertise through the process of capturing "lessons learned." Membership on this team should periodically rotate to prevent burnout and spread the travel burden. "Graduates" of the team bring their refreshed expertise and knowledge back to their home organizations.

- *Development of a set of design-for-manufacturing guidelines* (similar to those in Boothroyd, Dewhurst, & Knight, 2002) that are specific to the company and industry's products. This might be one of the tasks of the knowledge-capture team. Boothroyd et al.'s design-for-manufacturing guidelines are geared toward the automotive industry and are not readily applicable to biomedical products. Such guidelines are one mechanism for capturing manufacturing wisdom in a convenient and usable form for design engineers. However, since manufacturing processes are constantly being changed and improved, these guidelines need to be periodically updated.

- *Establishment of periodic cross-site training sessions* that allow design engineers to learn about the latest internal and external advances in manufacturing processes and manufacturing engineers to learn about new applications of products that may impact manufacturing choices. The knowledge-capture team may help to identify what topics might benefit each group. Training sessions at each site should be conducted by people from the other site. This will increase the number of personal cross-site contacts for both groups.

All of these recommendations require a significant investment of personnel and additional resources. However, given the large cost of reduced product yields and production delays, the investments are deemed highly likely to provide a worthwhile return. The author does not have information on whether these recommendations were followed, since management changed shortly after the recommendations were delivered.

9 Closing Thoughts

Much of the literature has focused primarily on interpersonal and communication issues between recognized members of the team. What the case study shows is that, at least in some cases, interpersonal issues within the team may be dwarfed by management and organizational issues external to the team. For example, the work priorities, resources, and incentives provided by the company influence whether people from separate sites *want* to work together in the first place and are willing and able to work to do so. In particular, global teams are often created and then expected to operate using work practices, resources, and management approaches similar to those used by face-to-face colocated teams. However, their needs are not the same. While organizations may know this in the abstract, they may still find unexpected surprises when a team becomes global. Informal practices that might not even be recognized as part of the work process, such as walking over to the manufacturing floor to discuss a problem or calling up a "friend" in manufacturing for his or her opinion, may fail to keep "working" once design and manufacturing groups are moved to geographically distant locations. Since many aspects that make a colocated team "work" are not formally recognized or understood, it can be very hard to diagnose what is no longer working in a global team.

This suggests that, in future work on global and virtual teams, the community of researchers and practitioners would be well served to further investigate not only the differences between colocated and global teams but also how organizational structures, including management, resources, and incentives, can be reconfigured to better support and encourage collaboration across sites.

Barriers to collaboration in virtual and global design teams are numerous, complex, and intertwined and arise from many aspects of the organization. In this study, many of the collaboration failures observed happened not for lack of communication tools but for lack of sufficient resources, accountability, incentives, and knowledge of cross-site personnel. An organization must use all of the above to create fertile ground for distance collaboration before it can flourish.

References

Barczak, G., & McDonough, E. F. (2003). *Managing global new product development teams* (Working Paper No. 03-001). Boston, MA: Institute for Global Innovation Management, Northeastern University.

Boothroyd, G., Dewhurst, P., & Knight, W. (2002). *Production design for manufacture and assembly.* New York: Marcel Dekker.

Chachere, J., Kunz, J., & Levitt, R. (2004). Can you accelerate your project using extreme collaboration? A model based analysis (Technical Report 154). Center for Integrated Facilities Engineering, Stanford University, Palo Alto, CA.

Daft, R. L., & Lengel, R. H. (1986). Organizational information requirements, media richness and structural design. *Management Science, 32*(5), 554–571.

Distefano, J. J., & Maznevski, M. L. (2000). Creating value with diverse teams in global management. *Organizational Dynamics, 29*(1), 45–63.

Handy, C. (1995). Trust and the virtual organization. *Harvard Business Review, 73*(3), 40–50.

Ishii, K. (1995). Life-cycle engineering design. *Journal of Mechanical Design, 117*, 42–47.

Kunz, J. C., Christiansen, T. R., Cohen, G. P., Jin, Y., & Levitt, R. E. (1998). The virtual design team. *Communications of the ACM, 41*(11), 84–91.

McDonough, E. F., Kahnb, K. B., & Barczaka, G. (2003). An investigation of the use of global, virtual and colocated new product development teams. *Journal of Product Innovation Management, 18*(2), 110–120.

McDonough, E. F., Kahn, K. B., & Griffin, A. (1999). Managing communication in global product development teams. *IEEE Transactions on Engineering Management, 46*(4), 375–386.

Sohlenius, G. (1992). Concurrent engineering. *CIRP Ann., 46*(4), 375–386.

Song, M., Berends, H., van der Bij, H., & Weggeman, M. (2007). The effect of IT and co-location on knowledge dissemination. *Journal of Product Innovation Management, 24*, 52–68.

Wickens, C. D., Lee, J. D., Liu, Y., & Becker, S. E. G. (Eds.). (2004). Research methods. In *An Introduction to Human Factors Engineering*. Upper Saddle River, NJ: Pearson Prentice Hall.

Index

Note: The *italicized f* or *t* following page numbers refers to figures and tables, respectively.